Iodine Made Simple

T0144528

Iodine Made Simple

Tatsuo Kaiho

CRC Press
Taylor & Francis Group
Boca Raton London New York

CRC Press is an imprint of the
Taylor & Francis Group, an **informa** business

CRC Press
Taylor & Francis Group
6000 Broken Sound Parkway NW, Suite 300
Boca Raton, FL 33487-2742

© 2017 by Taylor & Francis Group, LLC
CRC Press is an imprint of Taylor & Francis Group, an Informa business

No claim to original U.S. Government works

Printed on acid-free paper

International Standard Book Number-13: 978-1-1380-8077-5 (Hardback)
978-1-1380-6805-6 (Paperback)

Library of Congress Cataloging-in-Publication Data

Names: Kaiho, Tatsuo.
Title: Iodine made simple / Tatsuo Kaiho.
Description: Boca Raton : CRC Press, [2017] | Includes bibliographical references and index.
Identifiers: LCCN 2017009658| ISBN 9781138068056 (pbk. : alk. paper) | ISBN 9781138080775 (hardback (pod) : alk. paper) | ISBN 9781315158310 (ebook)
Subjects: LCSH: Iodine--Popular works. | Iodine--Industrial applications--Popular works. | Iodine--Therapeutic use--Popular works.
Classification: LCC TP245.I6 K37 2017 | DDC 661/.0734--dc23
LC record available at https://lccn.loc.gov/2017009658

Visit the Taylor & Francis Web site at
http://www.taylorandfrancis.com

and the CRC Press Web site at
http://www.crcpress.com

Contents

v

Preface

For better or worse, iodine became the focus of people's attention in the year 2011. On March 11, the Great East Japan Earthquake occurred and the subsequent accident at Fukushima No. 1 Nuclear Power Plant instantly made iodine the subject of media attention. However, it was not until sometime later that a distinction was made between stable iodine and radioactive iodine. The atomic weight of radioactive iodine is 131, whereas the atomic weight of stable iodine is 127. Incidentally, Japan is the world's second largest producer of iodine (stable iodine), producing approximately 30% of the world's supply. However, at the time this aspect did not attract people's attention.

Iodine was discovered from seaweed by Bernard Courtois in France in 1811. Exactly 200 years later, in the pivotal year of 2011, a symposium and seminar was held in iodine-producing Chiba under the sponsorship of the Iodine Industry Association and the Society of Iodine Science. Furthermore, a summary on iodine was published in chemical journals, and technical books were also published on iodine. However, these were too technical and difficult to understand by the general public.

With such a background, this book intends to explain the basic properties of iodine as well as various products and technology that use iodine, and was written in cooperation with iodine manufacturing technicians and university researchers.

When you hear the word "iodine" what do you imagine? Is it mouthwash or a disinfectant? If you are a science lover, maybe it is the iodine–starch reaction. In reality, iodine is widely seen in our daily life. In Chapter 1, we start by answering the question, "What is iodine?" In Chapter 2, we discuss "Iodine that exists around us," from commonly used mouthwash to other things that surprisingly contain iodine. Chapter 3 is a summary of "Iodine that sustains electronic and information materials." Then, in Chapter 4 under "Analysis using iodine," we list various analysis techniques implemented by university laboratories and corporate research institutions and factories that use iodine on a wide scale. Furthermore, in Chapter 5 titled "Innovative industrial technology starts with iodine," products and techniques applied in the industrial sector that take advantage of the properties unique to iodine are highlighted. In Chapter 6, "Iodine is a prerequisite in maintaining health," an explanation of iodine used in the medical sector, particularly in hospitals, is given. In Chapter 7, iodine and its relationship with vegetables and dairy products that fill our daily meals is discussed under "Iodine for vegetable production and livestock breeding." And finally, in Chapter 8, the future prospects that will be created by iodine are introduced.

I hope you, the reader, can deepen your understanding of iodine, which is a precious resource. In particular, I strongly encourage the younger generation to participate in the development of novel technologies using iodine.

In writing this book, I would like to express my deepest appreciation to the technicians of iodine manufacturers and iodine product manufacturers and to university researchers, all of whom cooperated in the writing of this book.

May 2017
Tatsuo Kaiho, Technical Advisor, PhD
Godo Shigen Co., Ltd.

Author

Tatsuo Kaiho is the technical advisor of Godo Shigen Co., Ltd., Chiba, Japan, one of the most well-known iodine manufacturing companies in the world. He is a former director of Nihon Tennen Gas Co., Ltd., Chiba, Japan, and a former director of Kanto Natural Gas Development Co., Ltd. The Vice Chairman of the Society of Iodine Science, Dr. Kaiho developed novel iodine-containing materials and processes and presented independent research at many conferences, including the International Conference of Hypervalent Iodine Chemistry (2010) and the Symposium of Iodine Science (2011). Dr. Kaiho has received several awards, such as the Distinguished Chemist Award from Chiba Prefectural Government (2001), the Organic Synthesis Award from Society of Synthetic Organic Chemistry Japan (2002), and the Society of Iodine Science Award (2012). He organized the 4th International Conference of Hypervalent Iodine Chemistry (2014) in Narita, Japan. Dr. Tatsuo Kaiho (born 1952) earned his MSc (1976) from Osaka University. He joined Mitsui Chemicals Co., Ltd. in 1976. On leave from Mitsui, he worked as a visiting scientist at Massachusetts Institute of Technology (1979–1982) and earned a PhD (1983) at Osaka University. From 2001 to 2003 he worked as a visiting professor at the Department of Engineering of Chiba University.

What is iodine?

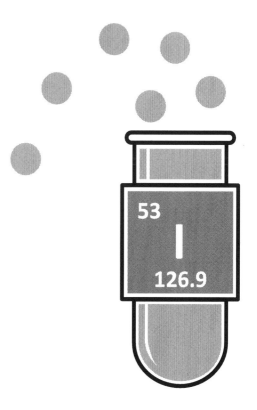

1

Iodine products that are deeply rooted in our everyday life

From mouthwash to LCD television

Iodine is deeply rooted in our everyday life. Products containing iodine atoms or molecules demonstrate properties such as "blocks X-rays, shows high reactivity, forms various compounds with most other elements, and provides strong anti-bacterial effect." These superior functions are seen in many areas in our daily life.

The specific capacity of iodine to absorb X-rays is used in contrast agents for X-rays. It is used in hospitals to create detailed images of the brain and heart. Iodine is also used to control light. It is used widely in LCD (liquid-crystal display) screens for TVs, mobile phones, and car navigation systems. Iodine is used in the polarizing film which acts as the switch on the screens. Depending on the specific direction the polarizing film arranges the iodine, it can filter light.

Industrial catalysts take advantage of the high reactivity of iodine. Acetic acid is produced from methanol and carbon monoxide as raw materials. This reaction is catalyzed by the metal (iridium or rhodium) complex with iodine and proceeds under mild conditions.

In automobile tires and airbags, nylon wire is used for reinforcement. To prevent oxidization or deterioration under high temperatures, potassium iodide or copper iodide is added as a stabilizer.

There is a recently developed insecticide in Japan called Flubendiamide, which combines fluorine with iodine. This chemical takes advantage of various halogen properties. Iodine molecules are high in antibacterial properties and have been used in a tincture of iodine for many years. Recently, various stabilized chemicals with a combination of polymers have appeared, a typical example being povidone iodine (Betadine).

Iodine is an essential element for living creatures. Iodine deficiency is caused by a lack of iodine intake. In Western countries, potassium iodate is added to table salt. Iodine is also indispensable for livestock and is provided as a feed additive [1].

Summary Box

- Without iodine, X-ray contrast agents and LCD polarizing films cannot be produced.

- Without iodine, humans and animals cannot live.

Characteristics of Iodine and Iodo-Compounds

① Biological Activity (Thyroid Hormone)
② Antibacterial and Antivirus Activity
③ X-ray Absorption Capacity
④ Inclusion and Complex-forming Ability
⑤ Electrical Conductivity Enhancement
⑥ Light Control Function
⑦ High Reactivity

Applications of Iodine

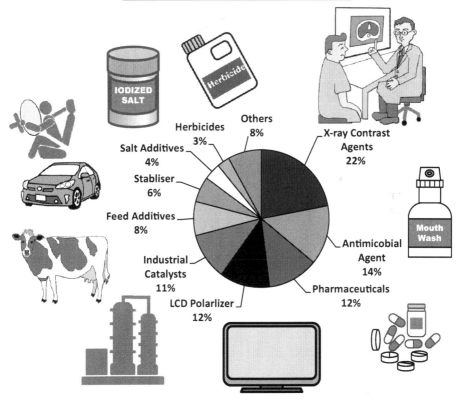

IODIZED SALT

Herbicide

Mouth Wash

Others 8%
Herbicides 3%
Salt Additives 4%
Stabliser 6%
Feed Additives 8%
Industrial Catalysts 11%
LCD Polarlizer 12%
Pharmaceuticals 12%
Antimicobial Agent 14%
X-ray Contrast Agents 22%

(Japan Iodine Industrial Association)

2

Iodine reacts with most elements

It is the heaviest and largest of all stable halogen elements

4

Iodine is one of the five halogen elements listed on the periodic table, and is located next to the noble gases. Of the stable halogens (fluorine, chlorine, bromine, iodine), excluding astatine which is radioactive, iodine is the largest element. A comparison of iodine with other halogen elements is shown in the chart, including van der Waals' radius which indicates the size of the atom, electronegativity which indicates the tendency of an atom to attract a bonding pair of electrons, and the binding energy for carbon. At room temperature, fluorine and chlorine are gas, and bromine is liquid. In contrast, iodine is a solid and has a melting point of 113.5°C.

The specific gravity of iodine is 4.93, which is heavy compared to metals (iron 7.85, titanium 4.61). Similar to camphor and naphthalene, iodine can sublimate, in other words change directly from solid to gas (without passing through the intermediate liquid phase). If iodine is placed in a jar, part of it change to a purple gas. Clarke number for iodine is 0.5, and is positioned as the 64th among the elements and is a vital resource. Iodine reacts to most elements excluding noble gases, and forms iodides. These properties of iodine are used in the recovery of rare metals.

Iodine is highly soluble in organic solvents. Depending on the characteristics of the solvent, solutions of varying color such as purple, red, and brown can be created. For example, chloroform and hexane solutions are purple, benzene, toluene, xylene solutions are red, and methanol, ethanol, and acetic acid solutions are brown. On the other hand, iodine does not dissolve well in water. However, in sodium iodide or potassium iodide solutions, the reaction shown in the bottom diagram occurs, producing polyiodide ion (I_3^-, I_5^-, etc.), which becomes easily soluble.

As described above, among the halogen elements, iodine is very unique [2].

Summary Box

- Iodine is a solid halogen element at room temperature with sublimabilty.

- Iodine has properties similar to metal.

Periodic table and Halogen Elements

1 H																	2 He
3 Li	4 Be											5 B	6 C	7 N	8 O	9 F	10 Ne
11 Na	12 Mg											13 Al	14 Si	15 P	16 S	17 Cl	18 Ar
19 K	20 Ca	21 Sc	22 Ti	23 V	24 Cr	25 Mn	26 Fe	27 Co	28 Ni	29 Cu	30 Zn	31 Ga	32 Ge	33 As	34 Se	35 Br	36 Kr
37 Rb	38 Sr	39 Y	40 Zr	41 Nb	42 Mo	43 Tc	44 Ru	45 Rh	46 Pd	47 Ag	48 Cd	49 In	50 Sn	51 Sb	52 Te	53 I	54 Xe
55 Cs	56 Ba	57~71 La~Lu	72 Hf	73 Ta	74 W	75 Re	76 Os	77 Ir	78 Pt	79 Au	80 Hg	81 Tl	82 Pb	83 Bi	84 Po	85 At	86 Rn
87 Fr	88 Ra	89~103 Ac~Lr	104 Rf	105 Db	106 Sg	107 Bh	108 Hs	109 Mt	110 Ds	111 Rg	112 Cn						

57 La	58 Ce	59 Pr	60 Nd	61 Rm	62 Sm	63 Eu	64 Gd	65 Tb	66 Dy	67 Ho	68 Er	69 Tm	70 Yb	71 Lu
89 Ac	90 Th	91 Pa	92 U	93 Np	94 Pu	95 Am	96 Cm	97 Bk	98 Cf	99 Es	100 Fm	101 Md	102 No	103 Lr

5

Comparison of the Halogen Elements

	Unit	Fluorine	Chlorine	Bromine	Iodine
Van der Waals' radius	pm	147	175	185	198
Ionic Radius	pm	133	181	195	216
Electronegativity	Allred-Rochow	4.1	2.83	2.74	2.21
Binding Energy	kcal/mol	115	83.7	72.1	57.6
Melting Point	°C	−219.62	−101.01	−7.2	113.5
Boiling Point	°C	−118.14	−33.97	58.78	184.3

Solubility of Iodine

Temp. (°C)	H_2O (g/l)	10% NaI (g/l)
20	0.293	9.6
30	0.340	10.3
40	0.549	10.9
50	0.769	11.7

▶ $NaI + I_2 \rightarrow NaI_3$

Glossary

Clarke Numbers : Numbers expressing the average content of the chemical elements in the earth's crust.

3

Iodine was discovered in France

6

Bernard Courtois, the chemist who discovered iodine

From ancient times, seaweed ash has been used as an ingredient for glass and fertilizers in the Brittany and Normandy regions of northwest France. In particular, during the Napoleonic Wars in the early nineteenth century, saltpeter (potassium nitrate, KNO_3), a component of gunpowder, was widely manufactured from seaweed, which is a souse of potassium. When manufacturing saltpeter, Bernard Courtois, a chemist, added too much sulfuric acid to seaweed ash, causing the iodine to evaporate and creating iodine crystals. To this day, the historic ruins of a stone seaweed incinerator can be seen in Brittany. Furthermore a monument to the discovery of iodine and buildings of the old iodine factory can be seen near the coast.

It was in 1811 that Courtois obtained those iodine crystals. Courtois, with the help of two friends (Charles-Bernard Desormes and Nicolas Clément) and Joseph Louis Gay-Lussac, verified iodine to be a new element and published an article in a French journal (*Annales de Chimie*) in December, 1813 [3a].

The Napoleonic Wars ended in 1815, and in 1820, the pharmacological properties of iodine were recognized, with Courtois starting commercial production of iodine. Courtois' contribution to the medical field was acknowledged and he received a Montyon Prize from the Dijon Academy in 1831.

Although Courtois died in 1838 without any assets, a dinner party was held on December 9, 1913 in Dijon, his birthplace, to praise his achievements regarding his discovery of iodine. A monumental plate was placed on the outer wall of his birthplace with the inscription, "Bernard Courtois, chemist who discovered iodine in 1811, was born in this house [3b]."

Approximately 200 years have passed since iodine was recognized as an element. And now, the third century of iodine chemistry has just begun.

Summary Box

- Iodine was discovered from seaweed ash.

- Two hundred years have passed since the discovery of iodine.

Birthplace of Iodine : France

Monument of Iodine Discovery
(See color insert.)

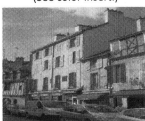

Courtois' home in Dijon, France
where Bernard Courtois was born
(See color insert.)

Plaque on the Courtois' home
(See color insert.)

Remains of a seaweed kiln on
the coast of Brittany
(See color insert.)

Brittany & Normandy

Paris

Dijon

France

> 304.   A N N A L E S
>
> *Découverte d'une substance nouvelle
> dans le Vareck.*
>
> Par M. B. Courtois (1).
>
> Les eaux-mères des lessives de Vareck
> contiennent en assez grande quantité une
> substance bien singulière et bien curieuse ;
> on l'en retire avec facilité : il suffit de verser
> de l'acide sulfurique sur ces eaux-mères et
> de chauffer le tout dans une cornue dont le
> bec est adapté à une alonge , et celle-ci à un
> ballon. La substance qui s'est précipitée
> sous la forme d'une poudre noire-brillante ,
> aussitôt après l'addition de l'acide sulfuri-
> que, s'élève en vapeur d'une superbe couleur
> violette quand elle éprouve la chaleur ; cette
> vapeur se condense dans l'alonge et dans le
> récipient , sous la forme de lames cristallines
> très-brillantes et d'un éclat égal à celui du
> plomb sulfuré cristallisé ; en lavant ces lames
> avec un peu d'eau distillée , on obtient la
> substance dans son état de pureté.
>
> (1) Cette découverte a été annoncée le 6 décembre,
> à la séance de la première classe de l'Institut, par
> M. Clément.

The front page of Courtois' historic
publication reporting discovery of iodine

Glossary

Kelp ash :The ash obtained by burning various brown seaweeds
such as Laminaria, Ecklonia, and Sargassum.

4

Iodine has been produced up to now by various methods using seaweed, oil brine, and natural gas brine as raw materials. The ashing method first began in nineteenth century France with the discovery of iodine. This method involves drying and burning seaweed collected from the seashore to make seaweed ash. Iodine concentration in seaweed ash is 0.5%–1.0%. The ash is lixiviated with water, which extracts the soluble salts, and the liquid is concentrated. Less soluble salts crystallize out and are removed. Sulfuric acid is then added to the liquid, and hydriodic acid is liberated. The liquid is run through an iodine still and gently warmed with oxidants, manganese dioxide, or chlorine. When the iodine distils over, it is collected.

In the copper method, a reaction is created by placing cuprous sulfate and ferrous sulfate in a brine. Copper iodide is allowed to precipitate, then any sediment is filtered and washed out in order to obtain crude copper iodide with approximately 50% iodine concentration. Next, after drying, the crude copper iodide is heated and oxidative decomposition is carried out to obtain iodine. In the activated carbon absorption method, sulfuric acid is added to the brine and after adjusting pH value to 2–4, sodium nitrite is added as an oxidant to liberate the iodine. Activated carbon, at approximately 7 times the quantity of free iodine, is added to absorb the iodine. Next, sodium hydroxide and sodium carbonate are added to the activated carbon with the absorbed iodine and heated, which is subsequently eluted as sodium iodide. This solution is concentrated in an iron pot, then acidified by sulfuric acid and oxidized with chlorine, resulting in the precipitation of iodine.

The starch method allows iodine to be absorbed by starch instead of activated carbon, and this method was used in the former Soviet Union and Japan for a time. Starch that is approximately 120–150 times the amount of free iodine is added to the free iodine brine. The absorbed iodine is extracted by water or sodium hydrogen sulfite. Starch does not absorb any other mineral elements and is thus considered to be ideal as an absorbent. However, the significantly large volume of starch and amount of chemicals required for production is problematic [4].

Summary Box

- Changes in the raw materials used for iodine production.

- From seaweed to brine.

Changes in the methods to produce iodine

Ashing, copper, activated carbon, and starch methods

Production Method in the Old Time

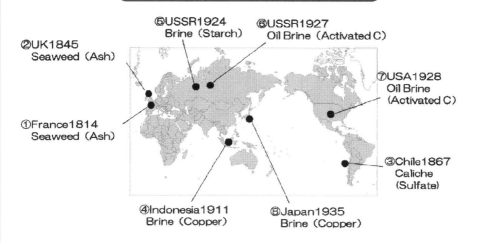

②UK1845
Seaweed (Ash)

①France1814
Seaweed (Ash)

⑤USSR1924
Brine (Starch)

⑥USSR1927
Oil Brine (Activated C)

⑦USA1928
Oil Brine
(Activated C)

③Chile1867
Caliche
(Sulfate)

④Indonesia1911
Brine (Copper)

⑧Japan1935
Brine (Copper)

Iodine Production Method and Reaction in the Old Time

Method	Reaction
Calcination Method	$2NaI+3H_2SO_4+MnO_2 \rightarrow I_2+2NaHSO_4+MnSO_4+2H_2O$ $2NaI+Cl_2 \rightarrow 2NaCl+I_2$
Copper Method	$2NaI+2CuSO_4+2FeSO_4 \rightarrow 2CuI+Na_2SO_4+Fe_2(SO_4)_3$ $2CuI+O_2 \rightarrow 2CuO+I_2$
Activated Carbon Method	$2NaI+2NaNO_2+2H_2SO_4 \rightarrow I_2+2NO+2H_2O+2Na_2SO_4$
Starch Method	$2NaI+NaClO+H_2SO_4 \rightarrow I_2+NaCl+Na_2SO_4+H_2O$

9

Iodine Purification Equipment

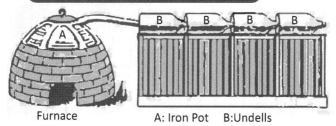

Furnace

A: Iron Pot   B:Undells

5

10

Although a country with few natural resources, Japan is proud to be the second largest producer of iodine following Chile. Hence, iodine is a precious resource for Japan. Although iodine exists in seawater, its concentration is low at 0.05 ppm and it is very difficult to extract economically, despite the progress in modern technology.

In Japan, iodine is produced from sodium iodide (100–150 ppm) found in underground brine which is pumped up with natural gas. Currently, domestic iodine production is 9,520 t (2013). Approximately 80% is produced in Chiba Prefecture and the remaining 20% comes from Niigata and Miyazaki Prefectures.

The world's largest iodine producer is Chile. Iodine is found in Chilean saltpeter (KNO_3) at a high concentration, approximately 400 ppm, in the form of iodates (IO_3) such as lautarite [$Ca(IO_3)_2$] and dietzeite [$Ca(IO_3)_2 \cdot CaCrO_4$], etc. (see Section 8). Iodine production in Chile was 20,100 t in 2013. As shown in the graph, production ratio by country is: Chile 60%, Japan 28%, and other countries, including the United States, Indonesia, Russia, and Azerbaijan, comprising the remaining 12%.

In the past, Japan was the top iodine producer in the world. However, due to concerns regarding ground subsidence, Japan enacted some self-imposed restrictions and production leveled off. On the other hand, Chile has continued to increase production to present levels (see chart) [5].

Iodine was discovered in 1811 in France from seaweed ash. To this day, some chemists think iodine is produced from seaweed. But in reality, iodine is land's gift, produced from underground brine or Chilean saltpeter.

Compared to concentration in brine, iodine concentration in seawater is extremely low, equivalent to one part per 2,000–3,000 parts. However, the volume of saltwater is inexhaustible. If an innovative concentration technique for seawater is developed, iodine may once again become a marine gift.

Summary Box

- Japan's iodine is produced in Chiba, Niigata, and Miyazaki.

- Iodine is collected from underground brine or Chilean saltpeter.

Iodine Production in Japan and in the World

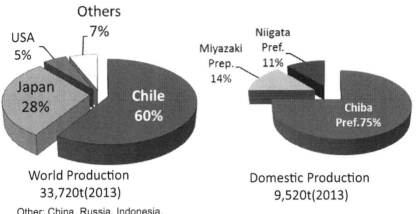

World Production
33,720t(2013)

Other: China, Russia, Indonesia,
Turkmenistan, Azerbaijan

Domestic Production
9,520t(2013)

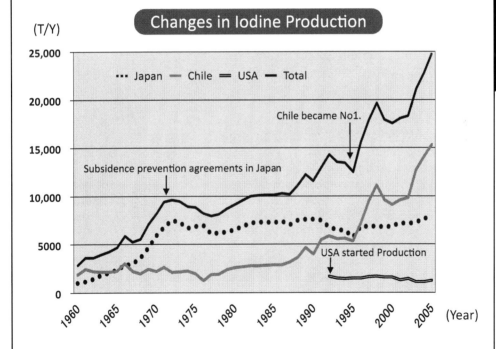

Changes in Iodine Production

(T/Y)

Chile became No1.

Subsidence prevention agreements in Japan

USA started Production

(Year)

Iodine is harvested along with natural gas

A gift from ancient times

Major water-soluble natural gas deposits in Japan are geographically distributed mainly in Chiba, Niigata, Miyazaki, and Okinawa Prefectures. In particular, development in Chiba is thriving. Called the Minami Kanto Gas Field, it has the greatest production and reserves deposits of all the water-soluble natural gas fields in Japan. The Minami Kanto Gas Field is a large water-soluble gas field centred in Chiba and covers a large area in Ibaraki, Saitama, Tokyo, and Kanagawa Prefectures [6a].

Natural gas produced from the Minami Kanto Gas Field is approximately 99% methane and this clean energy contains very little carbon monoxide, sulfur, etc. Due to its high calorific value and efficiency as an energy source, it is widely used for city gas, etc. Combustible natural gas is divided into wet natural gas, coal field gas, dry natural gas, etc., depending on its current state. Dry natural gas is methane originating from microbes dissolved in underground stratum water.

Natural gas is produced in Chiba is from the geological stratum called the upper total layer group. This stratum is made primarily of sandstone and mudstone which accumulated on the ocean floor during the geological age called the Quaternary Pleistocene Period (~400,000–3,000,000 years ago). Gas dissolved into the stratum water is deposited in the alternating layers of sandstone and mudstone, creating a gas layer.

This stratum water is called "brine" and is ancient seawater which has been locked up in this stratum. As a result, the components are very close to present seawater and are shown in the table. Stratum water characteristically contains approximately 2,000 times the iodide ions as that of seawater, at approximately 100 mg/L, and with very few sulfate ions [6b]. The origin of iodine is thought to be animal and plant debris which accumulated on the ocean floor in ancient times or algae with iodine deposits.

Brine is drawn up from 500 to 2,000 m underground along with natural gas, and isolated from the natural gas using a separator. Next, sand and other impurities are removed from the brine in a settling tank, and then the brine is sent to the iodine production process.

Summary Box

- Iodine is harvested from ancient seawater.

- The brine in Minami Kanto Gas field contains approximately 100 mg/L iodine.

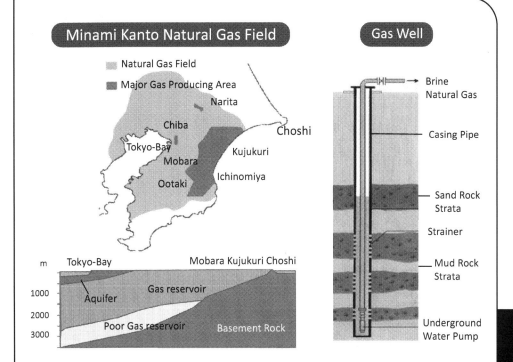

Minami Kanto Natural Gas Field

Natural Gas Field
Major Gas Producing Area

Narita
Chiba
Choshi
Tokyo-Bay
Kujukuri
Mobara
Ichinomiya
Ootaki

m   Tokyo-Bay   Mobara Kujukuri Choshi
1000   Aquifer   Gas reservoir
2000
3000   Poor Gas reservoir   Basement Rock

Gas Well

Brine
Natural Gas

Casing Pipe

Sand Rock Strata

Strainer

Mud Rock Strata

Underground Water Pump

Composition of Brine & Seawater

"Iodine concentrations in brine is 2,000 times higher than that in natural seawater."

Chemical	Ion	Chiba Brine (mg/L)	Sea water (mg/L)
I^-	iodide	100	0.05
Cl^-	Chloride	19,000	18,230
Br^-	Bromide	130	56
Na^+	Sodium	10,000	9,350
K^+	Potassium	300	356
Ca^{2+}	Calcium	190	290
Mg^{2+}	Magnesium	500	1,160
SO_4^{2+}	Sulfate	1~5	2,500
HCO_3^-	Carbonate	1,000	105
NH_4^+	Ammonium	200	1.5

pH	7.8	8.2

7

After natural gas is separated from brine which was drawn up from the ground 500–2,000 m beneath the surface, the brine is sent to the iodine production process. To extract iodine from brine, two methods are currently used, the blow-out method (upper diagram) and the ion-exchange resin method (middle diagram).

1. The blow-out method uses iodine's ability to easily vaporize and is suited for high-temperature brine treatment. The iodine-rich brine is acidified with an oxidizing agent, such as chlorine or sodium hypochlorite, to liberate iodine (I_2). The oxidized brine is then diffused to the top of a blow-out tower to vaporize the iodine. The dissipated iodine gas is reduced and concentrated at the same time into iodide ions (I^-) with an absorbent (sulfurous acid gas solution or sodium hydrogen sulfite). When the iodide ions in the absorbent are reoxidized with chlorine or sodium hypochlorite, it forms a muddy precipitation due to the high specific gravity of iodine. When heated and melted in a melting separator, the mud-like iodine descends to the lower layer due to its heavy specific gravity, and can be separated from the water. Finally, the melted iodine is either solidified and crushed into flakes using a flaker or made into small grains called prills, with a falling granulator, forming an iodine product. Today, most domestic and overseas manufacturers use this blow-out method.

2. In the ion-exchange resin method, iodide ion is partially oxidized with an oxidant, creating a polyiodine ion (I_3^-) state. This in turn is mixed with anion exchange resin, so the iodine can be obtained through absorption. In the melting process, the product is formed in the same refining process as the blow-out method. While the ion-exchange resin method is flexible in terms of volume, if the water temperature is high or if the brine contains many impurities, the efficiency of the process decreases [7]. This method of iodine production is only used at a limited number of factories in Japan.

Summary Box

- The blow-out method uses iodine's volatile properties.

- The ion-exchange resin method enables production at various scales.

Harvesting iodine from ancient seawater!

14

Iodine is volatile and easily absorbed

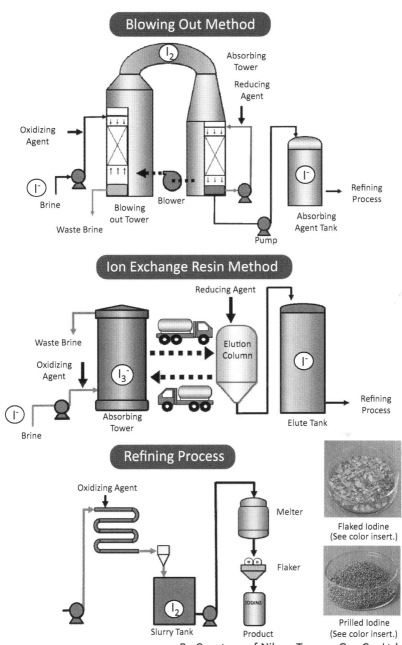

Blowing Out Method

I₂

Absorbing Tower

Reducing Agent

Oxidizing Agent

Brine

Blowing out Tower

Blower

Waste Brine

Pump

Absorbing Agent Tank

Refining Process

Ion Exchange Resin Method

Reducing Agent

Waste Brine

Oxidizing Agent

Brine

Absorbing Tower

Elution Column

Elute Tank

Refining Process

15

Refining Process

Oxidizing Agent

Melter

Flaked Iodine
(See color insert.)

Flaker

Slurry Tank

Product

Prilled Iodine
(See color insert.)

By Courtesy of Nihon Tennen Gas Co. Ltd.

Glossary

Ion Exchange Resin: Synthetic polymers containing positively or negatively charged sites that can attract an ion of opposite charge from a surrounding solution.

Iodine can be harvested from saltpeter in Chile

16

Crude ore caliche

There are many saltpeter deposits within the 700 km long, 15–80 km wide region of the Atacama Desert in northern Chile. This crude ore caliche contains nitrate at a 3%–12% concentration and iodate at a 0.04% concentration.

The production process of iodine is as follows: The crude ore is finely ground and nitrates and iodates are extracted with warm water, and then cooled so that sodium nitrate can be precipitated and separated. The process is repeated with the mother liquid, and through this crude ore extraction, 6–12 g/L iodate solution is obtained. To obtain the iodine, the iodic acid must be reduced. Part of the above solution is reduced using sulfur dioxide gas which is obtained by burning sulfur, creating an iodide ion solution.

Next, this iodide ion solution is mixed with the remaining iodic acid solution, with these acting alternately as an oxidant and a reducing agent, separating and precipitating the iodine (reaction formula). The precipitated iodine settles and is then sent to the refining process. The remaining iodine solution is collected using the blow-out method. Previously, the sublimation method was used in the refinement of crude iodine. However, iodine is now melted, refined, and formed into a granular product using the falling-type prilling method.

Chlorine is not used in Chile's iodine production method, and using iodine as an oxidant and reducing agent is said to be extremely efficient [8a,b].

The major obstacle in Chile's iodine production is the amount of water required for extraction. In desert areas, groundwater use is limited and transporting seawater more than 100 km from the coastline is difficult. Moreover, the processing plant is located on a highland 1,000–1,500 m above sea level and pumping up the water would be very costly.

However, in recent years, some manufacturers have installed seawater pipelines (see photo) from the ocean to the deposits.

Summary Box

- Iodine is harvested from the desert region in Chile.

- Chile's iodine production is efficient, and does not use chlorine.

Iodine Mine & Iodine Plant in Chile

(See color insert.)

Iodine Production Area in Chile

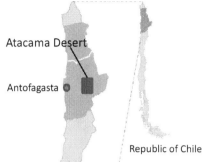

Atacama Desert

Antofagasta

Republic of Chile

Iodine Production Process in Chile

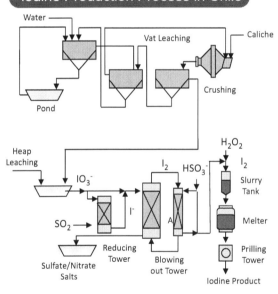

Water

Vat Leaching

Caliche

Crushing

Pond

Heap Leaching

H_2O_2

I_2 HSO_3^- I_2

IO_3^-

Slurry Tank

SO_2

I^-

A

Melter

Reducing Tower

Blowing out Tower

Prilling Tower

Sulfate/Nitrate Salts

Iodine Product

Pipe Line of Sea Water

Caliche Ore

Lautalite $(Ca(IO_3)_2)$
Dietzeite $(Ca(IO_3)_2 \cdot CaCrO_4)$
(See color insert.)

Reaction Formula

(1) $IO_3^- + 3SO_2 + 3H_2O \rightarrow I^- + 3SO_4^{2-} + 6H^+$

(2) $5I^- + IO_3^- + 6H^+ \rightarrow 3I_2 + 3H_2O$

Glossary

Caliche ore: Caliche is a sedimentaryrock, a hardened natural cement of carcium carbonate that binds other materials. Caliche ore in northern Chile contains nitrate and iodate.

Iodine is recycled

Iodine resources are limited

In Japan, iodine is produced from underground brine and is drawn up along with natural gas. However, due to concern over land subsidence from the excessive extraction of brine, the amount which can be pumped out is regulated. On the other hand, in Chile, iodine is produced by eluting iodate from Chilean saltpeter in areas such as the Atacama Desert. Difficulties in water procurement in the desert region limits production.

Against such a background, a rapid increase in iodine production is difficult. Therefore, various iodine manufacturers have begun efforts in iodine recovery and recycling. Some used solutions containing iodine have been recovered using various methods. Used iodine can be collected in various forms such as solutions, organic solvent solutions (polar, nonpolar), solids, slurry, etc.

During recovery, iodine content, form of iodine, and the existence of impurities such as heavy metals are determined by utilizing various analytical instruments. Next, a combination of recovery methods such as filtration, electrodialysis, high-temperature decomposition, and concentration are selected according to the form of the iodine.

Examples of high recovery rates include recovered liquids from polarizing films and recovered X-ray contrast agents. For inorganic iodine solutions, insoluble matter is first filtered out, then iodide is extracted using a dialysis membrane, and then concentrated. This process enables the iodine to return to the production stage. On the other hand, if an electrolytic membrane is used to change the dialysis membrane solution into a hydroiodic acid solution and then further concentrated and refined, raw material for inorganic iodine compounds can be obtained.

Waste liquids containing organic iodine compounds such as the X-ray contrast media solutions are first filtered of impurities, and then the recovery solution is decomposed at high temperature. The resultant iodine is absorbed by alkaline and recovered. The sodium iodide solution generated during the recovery process is returned to the iodine production stage, and is converted into an iodine product through the process outlined in Section 7. In this way, the precious iodine resource is recovered and recycled [9a,b].

Summary Box

- Inorganic iodine can be extracted and concentrated through electrodialysis.

- Organic iodine can be separated at high temperatures and recovered as inorganic iodine.

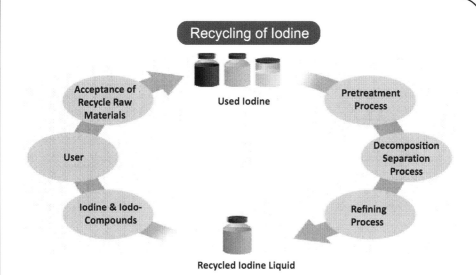

Recycling of Iodine

Used Iodine

Acceptance of Recycle Raw Materials → Pretreatment Process → Decomposition Separation Process → Refining Process → Iodine & Iodo-Compounds → User → Acceptance of Recycle Raw Materials

Recycled Iodine Liquid

Sample Status	Solution, organic solvents (polar, non-polar), solid materials, slurry
Iodine Content	Low (10 g / L or below) to High (100 g / L or over)
Iodine Form	I^-, I_2, IO_3^-, IO_4^-, aliphatic iodine compounds, aromatic iodine compounds, etc.

Electro-dialysis Method

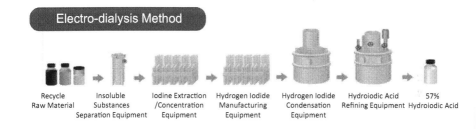

Recycle Raw Material → Insoluble Substances Separation Equipment → Iodine Extraction /Concentration Equipment → Hydrogen Iodide Manufacturing Equipment → Hydrogen Iodide Condensation Equipment → Hydroiodic Acid Refining Equipment → 57% Hydroiodic Acid

High Temperature Decomposition Method

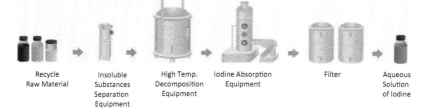

Recycle Raw Material → Insoluble Substances Separation Equipment → High Temp. Decomposition Equipment → Iodine Absorption Equipment → Filter → Aqueous Solution of Iodine

By Courtesy of Godo Shigen Co. Ltd.

10

Iodine circulates around the earth

Although distributed throughout the ocean, land, and air, the majority of iodine on the earth exists in seawater. However, iodine does not remain there. Actually, iodine circulates in the lithosphere, hydrosphere, and atmosphere, as shown in the diagram.

In the 1970s, James Lovelock, who advocated the Gaia hypothesis, used a gas chromatograph/electron capture detector (GCECD) and discovered that methyl iodide (CH_3I), a volatile organic iodine, is widely distributed in seawater and in the atmosphere. He showed that the inorganic iodine in seawater is organified (methylation) though biological reactions in macroalgae such as kelp, microalgae, and marine bacteria. As it evaporates into the atmosphere, it circulates dynamically throughout the global environment. Furthermore, other chemical forms of volatile organic iodine have been discovered, including diiodo-methane (CH_2I_2), chloroiodomethane (CH_2ClI), and ethyl iodide (CH_3CH_2I).

Through photolysis, evaporated organic iodine changes to inorganic iodine such as iodine (I_2), hypoiodous acid (HIO), and iodic acid (HIO_3) and precipitates to ground and oceans with rain, etc. Some of this is transferred from the soil to plants, from plants to animals (thyroid), and eventually returns to the ocean. In other words, evaporation of iodine is a vital process in supplying iodine from the ocean to the land, providing sustenance for the life of vertebrates on land, including humans. From continued research, iodine evaporation has also been found to occur in land environments, such as rice fields and peatland. In particular, rice functions to release methyl iodide into the atmosphere, and this alone supplies 4% of the iodine in the atmosphere. It is estimated that the annual evaporation rate of iodine on the global scale is 300,000–400,000 t.

According to recent studies, organic iodine is thought to be involved in processes related to global climate change such as the formation of cloud condensation nuclei and marine aerosols [10].

Atmospheric circulation of iodine

Summary Box

- Seaweed and rice release iodine into the atmosphere.

- Iodine falls to the land in rain and returns to the sea.

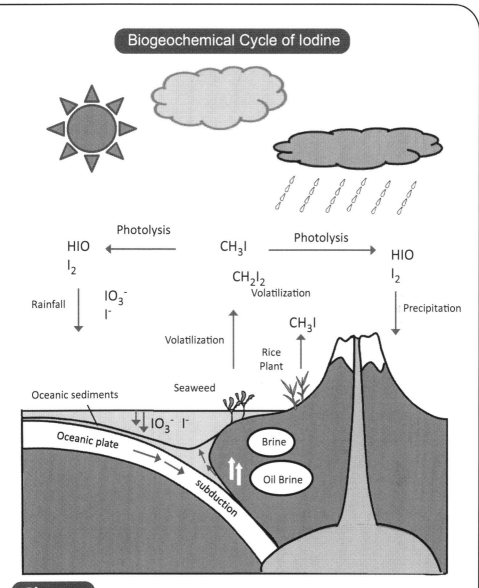

Biogeochemical Cycle of Iodine

Photolysis

HIO
I_2

CH$_3$I

Photolysis

HIO
I_2

IO_3^-
I^-

CH$_2$I$_2$

Volatilization

Rainfall

CH$_3$I

Precipitation

Volatilization

Rice
Plant

Seaweed

Oceanic sediments

IO_3^- I^-

Oceanic plate

Brine

Oil Brine

subduction

Glossary

Gaia theory: Gaia hypothesis proposes that all organisms and their inorganic surroundings on Earth are closely integrated to form a single and self-regulating complex system, maintaining the conditions for life on the planet.

In mountainous regions distant from the ocean, many cases of enlarged thyroid glands (goiter) have been recorded since ancient times. Seaweed was also known to be effective in treating such illnesses

The various medical benefits of seaweed were described in *Sennong Ben Cao Jing*, China's oldest medical book written around 2000 BCE. In the book, there is a reference to a type of seaweed thought to be brown algae *Sargassum fusiforme*, regarding a cure for tumors, typically goiter. *S. fusiforme*, a type of seaweed, is distributed throughout temperate and tropical zones and is considered to be the most evolved of all seaweeds.

On the other hand, many people who lived in the European Alps around the tenth century had enlarged thyroid glands. During the Middle Ages, there were many treatments for goiter, but by the thirteenth century, it is thought that ash from natural sponges was used.

However, it was not until the nineteenth century that it became clear that the active ingredient in the natural sponge was iodine.

Soon after the French chemist Courtois discovered iodine in 1811, Coindet, a doctor in Geneva, Switzerland, first used iodine in his medical practice in 1820. He made an alcohol solution from potassium iodide and iodine (tincture of iodine) and reported that this was tremendously effective in the treatment of goiter. His hypothesis was correct. However, patients who ingested the iodine tincture suffered severe stomachaches due to stomach lining irritation. As a result, this treatment did not become widespread.

Since then, it became increasingly clear that iodine is an essential element for vertebrates and plays a vital role in regulating metabolism as a thyroid hormone component.

Entering the twentieth century, potassium iodide and potassium iodate came to be widely used in the treatment of cretinism and goiter.

Shen-Nong
A legendary ancient Chinese god who tested various plants and identified medicinal substances.

2

Iodine around us

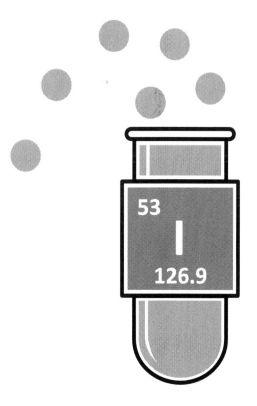

11

Iodine in mouthwash

Povidone iodine (typical brand "Betadine") is superior, not only in its bactericidal effect but also in its fast-acting properties. As a result, it is the principal disinfectant widely used in mouthwash, for sterilization of hands and fingers, and as an antiseptic for wounds. It is used widely in the clinical setting, from sterilization of wounds and washing of hands by doctors and nurses before starting surgery to sterilization during childbirth. It is also used in the common household, for gargling, in breath fresheners, handwashes, and to sterilize wounds, impetigo, or boils.

Povidone iodine is water soluble, binding iodine to its carrier povidone (PVP: polyvinyl pyrrolidone) (refer to the diagram). The amount of active iodine in 1 g of povidone iodine is 100 mg. Moreover, as an example, the concentration of active iodine in a 10% povidone iodine solution would be 1%. Povidone iodine is stable in the solution, but the free iodine is gradually released as the concentration of free iodine decreases in the solution [11a].

There are several theories regarding the mechanism of iodine. One is that H_2OI^+ generated by the reaction between iodine and water acts directly on the surface membrane protein of bacteria and viruses, resulting in a sterilizing effect.

When povidone iodine is used by persons with thyroid dysfunction, serum iodine cannot be regulated, which possibly affects thyroid hormone-related substances. In addition, shock and anaphylactoid symptoms (breathing difficulty, flushing, urticaria, etc.) due to hypersensitivity may occur. In the event such abnormalities occur, use should be immediately discontinued and the appropriate measures be taken [11b,c].

The National Aeronautics Space Administration (NASA) in the United States studied the efficacy of povidone iodine. There is an untold story that in the summer of 1969 when the spaceship Apollo 11 touched down in the Pacific Ocean after landing the first man on the moon, NASA decided to use povidone iodine to sterilize the spaceship which may have been contaminated from extraterrestrial bacteria, and to prevent contamination of the ocean [11d].

Povidone iodine ①

Summary Box

- Iodine is stabilized with povidone.

- Povidone Iodine is widely used as sterilizer in houses to hospitals.

Structure of Povidone Iodine

$$\left[\begin{array}{c} \overset{I_3^-}{\underset{O \cdots H^+ \cdots O}{\bigcirc}} \\ N \quad\quad N \\ -CH_2 - CH - CH_2 - CH - \end{array}\right]_n \left[\begin{array}{c} \bigcirc\!\!=\!\!O \\ N \\ -CH_2 - CH - \end{array}\right]_m$$

Gargle & Mouthwash

Povidone iodine gargle and mouthwash are suitable for sterilization, the washing of the throat and the bad breath removal.

(See color insert.)

Hand wash

Fingers and skin sterilization / disinfection

By Courtesy of Mundipharma Pte Ltd.

Sterilize the spacecraft

Apollo 11 splashed down in the Pacific Ocean in 1969.

By Courtesy of NASA.

Glossary

Povidone : Water-soluble polymer made from the monomer N-vinylpyrolidone.

12

26

Iodine has been known to have bactericidal effects since ancient times. However, due to its sublimability (the ability to change directly from the solid state to the vapor state), there were restrictions on its scope of use. Iodine absorption resin is a unique solid disinfectant which provides a solution to the above problem. Iodine absorption resin is characterized by an ionic bond between the basic anion ion-exchange resin and quaternary ammonium ion as a functional group (Amberlite [401S]). Under normal conditions, iodine is rarely released in water. When target ions, which have a higher tendency to ionize than iodine itself, approach the iodine absorption resin surface, force which induces iodine ion exchange is applied, resulting in the release of iodine ions. As a result, negative bacteria or viruses which come in contact with the iodine absorption resin surface are exterminated by the oxidative effect of the free iodine. Furthermore, as time passes, iodine penetrates the cell from the cell membrane and oxidizes inside of the cell. In this way, iodine absorption resin is able to instantly eliminate microbes such as bacteria, viruses, parasites, and mold. Specifically, iodine is effective against SARS (severe acute respiratory syndrome), influenza viruses, anthrax, Newcastle disease viruses, and *Aspergillus niger*, which seldom becomes drug resistant to iodine [12a].

In addition, iodine is effective against weaponized chemical gases, breaking down mustard gas, VX gas, etc. During times of war and chemical terrorism, iodine is used in gas masks [12b,c]. The shape of the iodine absorption resin is normally a prill bead approximately 0.5–1 mm. This is due to the fact that the larger the surface area which comes into contact with the microbes, the more effective the extermination. Iodine absorption resin can be used as a resin particle, or mixed with plastic (polypropylene, polyethylene, nylon, etc.) or fibers such as nonwoven fabric. It is used in a variety of ways in material used for sterilization of water and air containing bacteria and viruses (see Section 61).

Summary Box

- Ion-exchange resin can stabilize iodine.

- Iodine exterminates viruses, mold, and bacteria.

Iodinated Ion Exchange Resin

Positive charge of Amberlite resin attracts the negatively charged microorganisms.

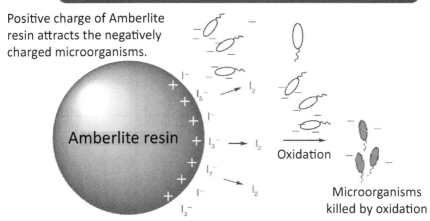

Oxidation

Microorganisms killed by oxidation

Positively charged surface structure of resin

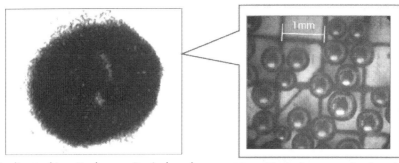

Iodinated Ion Exchange Resin beads

1mm

Magnified photograph of resin

Lethal Chemical Weapons Gas

VX gas

Mustard gas

Chemical Weapons Gases are extremely reactive and disabled by iodinated resin.

13

28

Iodine is effective on most molds and bacteria, as well as on various viruses such as influenza and AIDS (acquired immune deficiency syndrome). It is characterized by the absence of problems related to drug resistance experienced by antibiotics. However, iodine has certain disadvantages such as the tendency to sublime at room temperature (vaporize), a unique pungent odor, and the causing of skin irritations. However, such disadvantages can be overcome by combining iodine with amylose.

Amylose is found in natural starch at a concentration of approximately 20%, but there is no effective way of separating it. As a result, no progress has been seen in its industrial use. However, enzymatically synthesized amylose is a straight chain, unlike natural amylose. This amylose can be synthesized of any given molecular weight in a very narrow molecular weight distribution. These macromolecules are called bioamylose. Bioamylose has a spiral structure (host material) and captures other materials (guest materials) in its cavity (see the diagram). Furthermore, by creating "Amycel" by kneading amylose into rayon, the inclusion action of amylose can be provided to a fiber or sheet. Iodine included in amycel is effective for *Escherichia coli* O-157 and methicillin-resistant *Staphylococcus aureus* (MRSA).

By circulating air through the iodine including amycel filter, bacteria, and viruses in the air can be substantially inactivated. An odor-eliminating effect on malodorous substances such as ammonia and methyl mercaptan can also be observed. Iodine including Amycel consumes iodine as the functions are performed, and its dark purple color gradually fades. This is an indicator of the sustainability of the effect, and functions have a fixed period of performance. The color of the material can be used to gauge effectiveness of the functions. Iodine including Amycel is biodegradable and an environmentally circulating material [13].

Stabilizing iodine with enzymatically synthesized amylose

Summary Box

- The pungent odor of iodine is eliminated by amylose.

- Iodine-amycel is a biodegradable circulating material.

Structure of Amycel Iodine

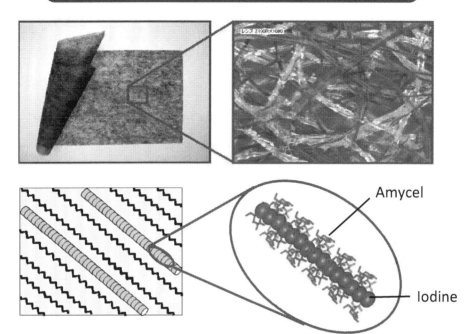

vvvv Rayon ⬭⬭⬭ Amylose Iodine (See color insert.)

Amycel

Iodine

Antivirus Activity of Amycel Iodine

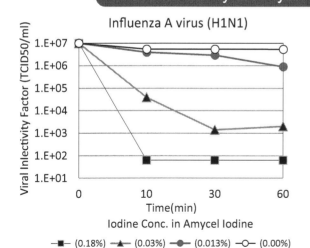

Influenza A virus (H1N1)

Viral Infectivity Factor (TCID50/ml)

Time(min)

Iodine Conc. in Amycel Iodine

■ (0.18%) ▲ (0.03%) ● (0.013%) ○ (0.00%)

Amycel Iodine Mask
(See color insert.)

By Courtesy of Ezaki Glico, Ohmi-Kenshi, Kanto Natural Gas Dev.

14

30

Cyclodextrin captures iodine

Cycloamylose stabilized iodine

Cyclodextrin can incorporate various substances in its molecule, and is used in various fields such as food, pharmaceutical products, and cosmetics.

For example, in the field of food, cyclodextrin is used to suppress bitterness in tea and various extracts (masking of bitter substances). In the field of pharmaceutical products, it is used to facilitate the dissolving of drugs and to enhance absorption efficiency (solubilization). Moreover, in the field of cosmetics, it is used to extend the life of perfumes (sustained release as the perfume is gradually released).

Cyclodextrin is an oligosaccharide of 6–8 continuous circular loops of glucose and is called a cyclic oligosaccharide (cycloamylose). Cyclodextrin molecules have a cylindrical shape with the edge of one side slightly narrowed, resembling a bucket. The inside of the molecule is hydrophobic and the outside is hydrophilic. The internal cavity has the ability to capture other molecules.

Cyclodextrin is added to iodine water which has been adjusted from iodine and potassium iodide, then heated, mixed, and left at room temperature to produce a dark brown iodine inclusion complex. By repeated filtering and washing, excess iodine and potassium iodide are removed, and when dried, a powder form of cyclodextrin iodine inclusion complex is obtained.

There is a correlation between the stability of cyclodextrin iodine inclusion complex (CDI) and the size of the cyclodextrin cavity, and can be seen in the following order: γ-CD>β-CD>α-CD. In particular, medium-size β cyclodextrin iodine inclusion complex (BCDI) was shown to have optimum stability and sustained iodine release. In this way, a stable iodine inclusion complex can be industrially manufactured with β-CD.

BCDI can be used for antibacterial films, food containers, medical equipment, filters, hoses, mats, fiber, and other resin mold products, and germicidal water treatment agents [14].

Summary Box

- β cyclodextrin iodine provides stability.

- Germicidal materials can be produced.

Cyclodextrin Cavity Size

	α-CD	β-CD	γ-CD
Molecular Weight	972	1135	1297
No of Glucose Units	6	7	8
Cavity diameter (Å)	4.7	6.0	7.5
Cavity height (Å)	7.9	7.9	7.9

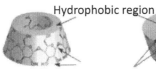

Hydrophobic region

Hydrophilic region

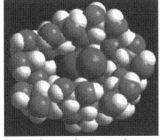

Cyclodextrin Inclusion Complexes with Iodine. Iodine is in the center of Cyclodextrin.

(See color insert.)

Stability of Cyclodextrin Inclusion Complexes with Iodine

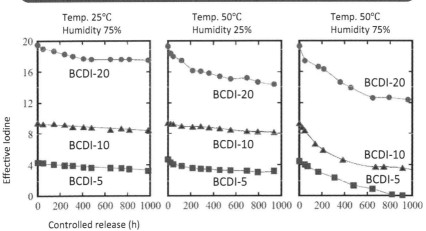

Temp. 25°C
Humidity 75%

Temp. 50°C
Humidity 25%

Temp. 50°C
Humidity 75%

By Courtesy of Nippoh Chemicals Co. Ltd.

15

Iodine compounds for mold extermination

Protecting cosmetics from microbial contamination is essential in order to maintain product quality. Although cosmetics are not required to be aseptic under the Pharmaceutical Affairs Law, cosmetics manufacturers generally strive for no viable bacterial count in their cosmetics.

Primary contamination at the manufacturing plant can be prevented by proper management of production and filling procedures. However, in order to prevent secondary contamination which occurs at the hand of the consumer, preservative ability is the most effective means. Therefore, while at the same time not having any viable bacterial count in the products is ideal, a preservative that is safe and effective with minimal quantity is sought.

Iodopropynyl butylcarbamate (IPBC) has poor solubility and a melting point of 64–68°C, but is an iodine-based fungicide which dissolves in most organic solvents. IPBC was developed in the United States in 1972. Its superior fungicidal properties are utilized in various fields such as wood preservation, building materials, sealants, adhesives, paints, surface preparation agents, fluids for metalwork, antibacterial agents for plastic and fiber, etc. In general, carbamate compounds have properties which inhibit blood cholinesterase activity. However, IPBC does not exhibit such action and is extremely safe (see the diagram). In 1996, IPBC was approved for use as a cosmetic preservative [15a]. The table indicates IPBC's effect on bacteria and fungi in comparison to parabens. IPBC's minimum inhibitory concentration is significantly lower than that of parabens, and a significantly superior fungicidal effect is obtained with a small dose. IPBC is high in fungicidal effect and has a high safety level at the same time.

Another iodine-based fungicide is diiodomethyl-p-tolysulfone. This compound is yellowish brown and comes in a fine powder. It is sold under the tradename "Amical 48." As shown in the diagram, it has a high safety level and is widely used as an antifungal agent for wood and paint [15b].

Iodopropynyl butylcarbamate diiodomethyl-p-tolysulfone

32

Summary Box

- Iodine-based fungicide with a high safety level.

- From cosmetics to paint and building materials.

Fungicide（IPBC）

I———≡≡≡———$CH_2OCONHBu\text{-}n$

IPBC

mp 64～68°C

Solubility 156mg (H_2O, 20°C)

Paint
Clothing
Cosmetics

LD_{50}=1100mg/kg (rat)

Biological Activity of IPBC

Type of Fungus	Parabens				IPBC
	Methyl	Ethyl	Propyl	Butyl	
Aspergillus niger	1000	400	200	200	1
Bacillus subtilis	2000	1000	250	125	500
Candida Fungus	1000	1000	125	125	6
Chaetomium	500	250	63	32	5
Escherichia coli	2000	1000	1000	4000	100
Pseudomonas aeruginosa	2000	1000	800	200	250
Staphylococcus aureus	4000	1000	500	125	100

MIC；minimum inhibitory concentration（ppm）

Fungicide（DMTS）

H_3C———◯———SO_2CHI_2

DMTS

mp 147～150 °C

Building materials
Leather

LD_{50} ： 9400mg/kg

Solubility of DMTS

Solvent	g/L
Water	0.0001
Isopropyl alcohol	10
Mineral Oil	＜4.0
Toluene	43
Ethanol	20
Ethylene glycol	10
Hexane	2
Xylene	33
n-Propyl acetate	263
Methyl ethyl ketone	25
N-Methyl-2-pyrrodinenone	33
Monoethanolamine	20

Biological Activity of DMTS

Type of Fungus	DMTS
Aspergillus niger	0.4
Chaetomium	0.2
Aspergillus oryzae	1.6
Fusarium (oxysporum)	6.2
Penicillium (Shitoriumu)	0.8
Bacillus subtilis	10
Staphylococcus aureus	6
Pseudomonas aeruginosa	＞1000
Sarumorera fungus	100

MIC；minimum inhibitory concentration（ppm）

Glossary

A carbamate is an organic compounds derived from carbamic acid (NH_2COOH).

16

34

Iodine as food coloring

Food Red Nos. 3 and 105

Food coloring stimulates one's appetite and enhances taste and increases enjoyment. Food coloring is roughly divided into three colors, namely red, yellow, and blue. Artificial food coloring, Red Nos. 3 and 105 contain iodine.

Red No. 3 is made of erythrosine and is categorized as edible tar dye. Its structure is shown in the diagram, and has such properties as a large molecular weight of 898, shows high heat resistance, and easily binds to protein. Industrially, it is produced by iodination reaction in fluorescein. It is mainly used to color food and industrial products. It is used as food coloring for fukujinzuke (sliced vegetables pickled in soy sauce), kamaboko (boiled fish paste), and cherries.

Another artificial red food coloring containing iodine is rose Bengal. Sodium salt is designated as a food additive for food Red No. 105. The structure of rose Bengal is created by replacing eight hydrogen molecules with four chlorine and four iodine molecules, and has a large molecular weight of 1,018. Potassium salt is also used as food coloring. It does not hold up well to sunlight and is unstable in acid. However, it has high heat resistance and good reducibility, and can be used in fermented foods and baked confectioneries. It is mainly used in kamaboko, Naruto rolls (spiral kamaboko), sausages, etc.

Some countries have banned its use due to concern regarding genetic damage. However, comparatively advanced toxicity tests in Japan have verified safety and its use has been approved.

Such artificial coloring is permitted in pharmaceutical products, quasi-drugs, and cosmetics. Rose Bengal is used as a plaque identifying agent in dentistry. In addition, rose Bengal is applied to the eye when carrying out the rose Bengal test, which is part of a special treatment by ophthalmologists, in order to test for conjunctive epithelium disorder. Other uses include use in cosmetics, including foundation, lipstick, and eye shadow [16].

Summary Box

- Coloring that shows high heat resistance and easily binds with protein.

- Coloring that is used for nonfood purposes such as cosmetics and pharmaceutical products.

Dyes Containing Iodine

Erythrosine

Rose Bengal

Cosmetic Use

Eye Shadow

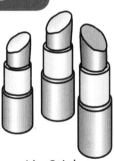

Lip Stick

Medical Use

Staining in Ophthalmology

Dental plaque staining

17

Iodine is essential for humans and animals

36

Thyroid hormone

Iodine is an essential microdose element for the living body and is indispensable for the life and growth of humans and animals. The human body contains approximately 20–30 mg of iodine. Daily intake of 0.090–0.25 mg of iodine is required, and half of the iodine in the body accumulates in the thyroid glands located under the chin (neck) (see lower left diagram), and constitute the materials used to produce thyroid hormones such as thyroxine and triiodothyronine (middle diagram) [17a]. These hormones stimulate the sympathetic nerves and promote metabolism of lipids and carbohydrates.

When thyroid hormones are insufficient, the pituitary gland increases secretion of thyroid-stimulating hormones in order to increase thyroid gland function, resulting in an enlarged thyroid or goiter. In addition, when thyroid hormones are insufficient, the result is decreased metabolism and physical strength, as well as impeded growth and mental retardation. In particular, pregnant women need to be concerned about iodine deficiency. When the mother has insufficient iodine, the fetus may experience development disorders and growth impediment of the cerebral and nervous system, in addition to increased risk of cretinism. These symptoms are collectively referred to as iodine deficiency.

Very few Japanese people show these symptoms as they take in the sufficient amount of iodine from seafood such as seaweed and sea fish. However, approximately 1.6 billion people throughout the world are at risk of developing iodine deficiency, which is considered to be a serious problem. Primarily, people living in mountainous areas such as the Alps, the Himalayas, and the Andes where the surface soil is stripped due to glaciers, and areas where the topsoil has been washed away due to frequent floods such as Bangladesh, are prone to goiter. As a main countermeasure, iodine compounds are added to salt to supplement iodine intake. Chiba Prefecture, which produces the majority of the iodine in Japan, cooperates with the Iodine Industry Association, and carries out iodine support operations for countries suffering from iodine deficiency such as Mongolia, Cambodia, and Sri Lanka [17b,c].

Summary Box

- Thyroid hormones are produced from iodine.

- Iodine deficiency results in enlarged thyroid and goiter.

Synthesis of Thyroid Hormones

Essential Amino Acid

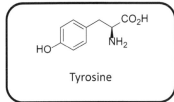

Tyrosine

Iodination →

Thyroid Hormone Precursor

X=H Monoiodotyrosine
X=I Diiodotyrosine

Thyroid Hormone

Triiodothyronine T3

Thyroxine T4

Action of Thyroid Hormones

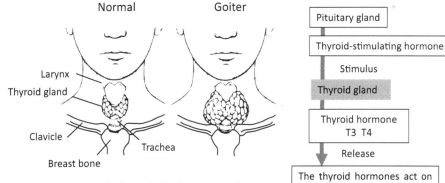

Normal   Goiter

Larynx
Thyroid gland
Clavicle
Trachea
Breast bone

Severe iodine defficiency results in thyroid enlargement (goiter).

Pituitary gland

Thyroid-stimulating hormone

Stimulus

Thyroid gland

Thyroid hormone
T3 T4

Release

The thyroid hormones act on nearly every cell in the body.

18

38

Why is iodine concentrated in seaweed?

Since the Japanese diet consists of seaweed and seafood, iodine intake in Japan is sufficient. As a result, there are very few cases of iodine deficiency among Japanese. On the other hand, there are cases exhibiting symptoms of iodine excess in some coastal areas of Hokkaido.

The concentration of iodine in seawater is low, approximately 0.05 mg/L. However, seaweed takes that iodine and condenses it 1000–10,000 times.

The majority of iodine contained in seaweed accumulates as inorganic iodine compounds, in such forms as iodide ion, triiodide ion, and iodate ion. Other rare forms of iodine have also been confirmed, such as thyroid hormone-like organic iodide compounds including diiodotyrosine.

According to a study analyzing iodine in seaweed, the ratio of iodide ions is 66% in *Sargassum* and 88% in kelp. The ratio of iodate ions is 29% in *Sargassum* and 10% in kelp. The ratio of organic iodine is 4.5% in *Sargassum* and 1.4% in kelp.

Additionally, the iodine concentration in edible seaweed is approximately 2,500 mg/kg in kombu, 140 mg/kg in wakame, and 40 mg/kg in nori (laver).

In this way, iodine abundantly accumulates in seaweed. However, how and why this accumulation takes place has remained a mystery.

However, according to a recent study by Professor Kuepper et al. of the University of Aberdeen in the United Kingdom, iodide ions in brown algae such as kombu and *Ecklonia cava* are oxidized to hypoiodous acid (HIO) by means of iodinated oxidase (V-IPO) on the cell membrane side, enabling free passage through the cell membrane, which is reduced back to iodide ions and the iodide ions accumulate within the cell. Furthermore, regarding the physiological significance of iodine in brown algae, the functions of iodine, such as its role as an antioxidant agent which detoxifies active oxygen including ozone and hydrogen peroxide, have been advocated [18].

Seaweed uses iodine for protection against oxidation

Summary Box

- Iodine is oxidized by an enzyme and passes through the cell membrane of seaweed.

- Iodine functions an antioxidant for seaweeds.

Iodine Concentration in Food

Animal Food	Iodine (μg/100g)	Vegetable Food	Iodine (μg/100g)
Sardines	268	Kelp	131000
Mackerel	248	Wakame	7790
Bonito	198	Porphyra	6100
Butter	62	Soybean	79
Chicken	49.9	Red beans	54
Egg (egg yolk)	48	Rice (milled rice)	39
Salmon	31.2	Greenpeace (raw)	20
Beef	16.4	Bread	17
Pork	17.8	Sweet potato	9.3
Normal milk	6	Onion	8.4

Iodine Uptake Mechanism

Brown Algae

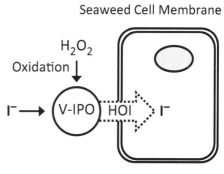

Seaweed Cell Membrane

H_2O_2

Oxidation

$I^- \rightarrow$ V-IPO HOI $\rightarrow I^-$

V-IPO : Iodide Peroxidase

Herapathite, also known as iodoquinine sulfate, was discovered and subsequently studied by toxicologist William Bird Herapath. In 1852, Herapath discovered that needle-like crystals had formed in the urine of a dog that was fed quinine (an antimalarial drug), into which iodine was accidentally dropped.

Herapath discovered that the overlapping crystals absorbed polarized light along a particular axis and appeared black in color.

Later, Polaroid Corporation, an American company, developed and marketed an artificial polarizing plate under the trade name Polaroid which embedded calcite, iodine, and herapathite (acidic sulfate quinine and acidic periodic acid) into an acetyl cellulose sheet, aligning the crystals. Through this invention, a large polarizing film could be obtained inexpensively, replacing the costly Nicol prism.

Polarizing film has many uses, such as polarizing filters and sunglasses. However, with the spread of LCD in recent years, demand has further increased.

While herapathite is a useful crystalized substance, the crystal structure, namely the precise atomic arrangement within the crystal lattice, was not clearly understood until recently.

Then, in 2009, the longtime mystery regarding herapathite was revealed. Professor Bart Kahr et al. could identify the structure. Herapathite was found to consist of triiodide ions aligned in a zigzag chain.

This structure is remarkably similar to that of the modern polarizing film created from triiodide ions and polyvinyl alcohol.

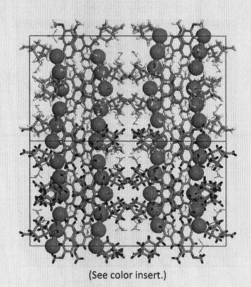

(See color insert.)

Herapathite
The crystal structure of herapathite. The gray spheres are iodine atoms.

Iodine that sustains electronic and information materials

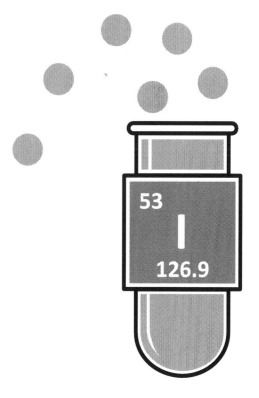

53

I

126.9

19

42

Using iodine to develop next-generation semiconductors

New etching gas, iodide trifluoromethane

Iodide trifluoromethane (CF_3I) has gained attention as etching gas in substitute of Freon for next-generation semiconductors.

When processing the silicon semiconductor integrated circuit, the circuit pattern is transferred to the photoresist, and after exposure, the exposed areas are etched. Then, metal used as metal wiring is embedded in the groove created by the etching. With half the distance of the center distance of the adjacent metal wiring as standard, the expression 65 nm process is used in reference to silicon semiconductors. The standard leading edge is 65 nm process, but the next generation is said to be 45 nm process, indicating a high integration trend.

Compared to C_4F_6 and CF_4, CF_3I has superior characteristics in the etching process under plasma conditions, such rarely releasing F radicals which cause damage to the photoresist, extremely low density, and relatively low ultraviolet ray (UV) intensity. On the other hand, concern that CF_3I-derived iodine may cause a negative effect on semiconductor or semiconductor-producing devices has been raised. However, even with the use of CF_3I, no iodine residue has been found on the processing film or the processing equipment surface, and subsequently no damage either.

The global warming potential (GWP) for conventional Freon etching gases is exceedingly high at 6,500 for CF_4 and 8,700 for C_4F_8. However, the value for CF_3I gas is 0.4, demonstrating an extremely environmentally friendly nature (see the diagram).

As a substitute material for Freon, CF_3I can also be used as a blower, acting as a gas jet to blow away dust, and has already been commercialized. As conventional products used LPG (liquefied petroleum gas) as the propellant, many blowers are combustible. However, CF_3I is noncombustible. This safety feature as a blower is fully utilized. In addition, since iodine is a heavy element, CF_3I was found to be more effective at thoroughly blowing away dust [19].

Summary Box

- Photoresist-friendly etching gas.

- An environmentally friendly etching gas with low GWP.

Etching by CF₃I

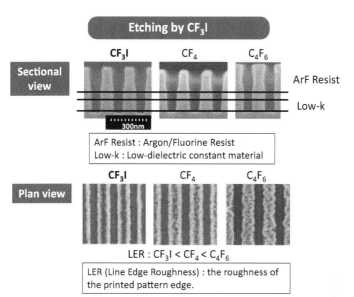

Sectional view — CF₃I, CF₄, C₄F₆ — ArF Resist, Low-k

300nm

ArF Resist : Argon/Fluorine Resist
Low-k : Low-dielectric constant material

Plan view — CF₃I, CF₄, C₄F₆

LER : CF₃I < CF₄ < C₄F₆

LER (Line Edge Roughness) : the roughness of
the printed pattern edge.

By Courtesy of New Energy and Industrial Technology Development Organization (NEDO).

Advantages of CF₃I Etching

Low fluorine radical density
LER improvement
Low GWP

Global Warming Parameters (GWP)

SF_6	: 23900
C_4F_8	: 8700
CF_4	: 6500
C_4F_6	: 290
C_5F_8	: 90
CF_3I	**: 0.4**
CO_2	: 1

Dry Etching by CF₃I

Damascene process: 1) Etching holes and trenches, 2) Electrodeposition of copper, 3) Planarization by Chemical Mechanical Polishing (CMP)

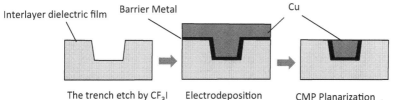

Interlayer dielectric film   Barrier Metal   Cu

The trench etch by CF₃I   Electrodeposition   CMP Planarization

Glossary

Global warming potential (GWP): GWPs are based on the heat-absorbing ability of each gas relative to that of carbon dioxide (whose GWP is standardized to 1).

20

Iodine in the manufacture of LCD projectors

Dry etching using high-purity hydrogen iodide gas

Among LCD (Liquid-Crystal Display) projector components, transparent pixel electrodes perform one of the most important functions. In an LCD, when voltage is applied to the transparent pixel electrodes, the orientation state of the liquid crystal changes, controlling the transmitted light. These transparent pixel electrodes use ITO (indium tin oxide) due to its high conductivity. ITO generally has an indium/tin ratio of 90/10, but is one of the most difficult materials to etch.

Traditionally, the wet method has been used to etch ITO. Chemical solutions with strong acidity such as aqua regia (hydrochloric acid/nitric acid mixture) are used. However, problems such as poor uniformity and instability, penetration of the etching fluid, difficulty in fine processing, etc., exist. To solve these problems, the dry etching method was developed. In the dry etching method, radicals and ions with high reactivity are generated in plasma and are applied to the thin film surface, creating a chemical reaction with the substrate surface material. Thereafter, these are converted to a compound using high vapor pressure and removed from the thin film.

Use of halogen gases such as chlorine, hydrogen chloride, and hydrogen bromide was attempted in the dry etching method. However, in comparison to these gases, hydrogen iodide gas exhibited a far superior performance.

Next, the ITO dry etching chemical reaction formula using hydrogen iodide gas was developed, as shown in the diagram. Indium iodides and tin iodides generated from the ITO reaction have a higher vapor pressure compared to corresponding chloride and bromide (see the diagram), and are easily removed from the thin film. Hydrogen iodide has an etching speed of 1,000 Å/minute and has superior pattern controllability.

When ITO thin film etching is observed by electron micrography, the etching surface for the most part was vertically straight with sharp corners and edges. In addition, the etched surface did not have a residue, proving to be an outstanding method [20].

Summary Box

- Metal iodides have high vapor pressure.

- Sharp corners and edges can be achieved when etching with iodine compounds.

High-grade Hydrogen iodine gas

Chemical Name	Hydrogen Iodide
Molecular Formula	HI
Molecular Weight	127.9
Property	Colorless Gas
Specific Gravity	4.4
Melting Point	-50.8°C
Boiling Point	-35.4°C
Purity	99.999%
CAS No	10034-85-2

Reaction of ITO & HI

$$In_2O_3 + 6HI \rightarrow 2InI_3 + 3H_2O$$
$$SnO_2 + 4HI \rightarrow SnI_4 + 2H_2O$$

Indium·Tin Halide Vapor Pressure

Vapor Pressure of Iodide is higher than bromide or chloride, and easily volatilizes.

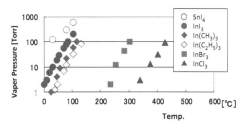

Comparison of Etching Performance

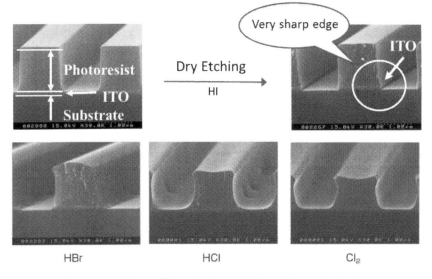

Very sharp edge

ITO

Photoresist
ITO
Substrate

Dry Etching
HI

HBr   HCl   Cl$_2$

By Courtesy of Godo Shigen Co. Ltd. and Mitsui Chemicals Co. Ltd.

45

21

46

Polarizing film (film with polarizing performance) is made primarily using iodine or dye to stain polyvinyl alcohol (PVA). The film is stretched and oriented to allow only polarized light in a fixed vibration direction to pass through.

In particular, iodine-type polarizing film has superior polarizing performance compared to dye-type polarizing film, and is used widely in LCD televisions and mobile phone screens. However, under high-temperature conditions, the dye type is considered to be more effective.

Iodine-type polarizing film used in LCDs is produced as follows: First, a PVA film which has expanded in water is soaked in an iodine/ potassium iodide solution. Next, it is treated with boric acid, and then stretched, rolled, and dried. As a result, a polarizing film with polyiodide ions (I_3^-, I_5^-) neatly aligned between the PVA molecule chain in the stretching direction is created.

Examining this film in detail, the polyiodide ions (I_3^-, I_5^-) are found to be sandwiched between the surface of the crystal part of the PVA with a polymerization degree of 2,400 and the noncrystal part of the PVA molecule chain, and the structure of both ends is closed and fixed with boric acid crosslinking (middle diagram).

The LCD principle is shown in the diagram. When a twisted liquid crystal is inserted between two sheets of polarizing filters, the polarizing direction of the filters intersects at right angles so light does not pass through. Light coming in from below is twisted 90° along the gap of the liquid crystal molecules, allowing light to pass through the filter above (passage of light). When voltage is applied, the liquid crystal molecules are aligned and untwist. Light coming in from above travels upward and cannot pass through the filter above (light is blocked). In other words, voltage is used as the trigger, and the liquid crystal assumes the role of an optical shutter.

This iodine-type polarizing film is essential in modern information equipment [21].

Summary Box

- Iodine-type polarizing film has superior polarizing performance.

- Polarizing film has various uses, from LCD televisions and mobile phones to cameras.

Production method of polarizer

| PVA Film | Swelling | Dyeing | Cross-Linking Streching | Polarizer |

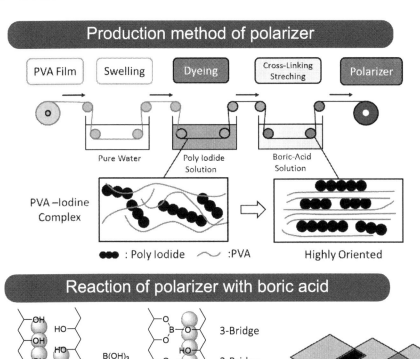

Pure Water — Poly Iodide Solution — Boric-Acid Solution

PVA –Iodine Complex

●●● : Poly Iodide ～ :PVA

Highly Oriented

Reaction of polarizer with boric acid

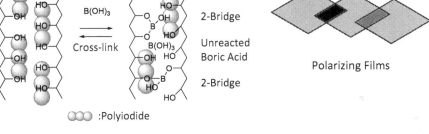

$B(OH)_3$

Cross-link

3-Bridge

2-Bridge

Unreacted
Boric Acid
$B(OH)_3$

2-Bridge

◯◯◯ :Polyiodide

Polarizing Films

47

LCD Panel

Application of Polarizing Film

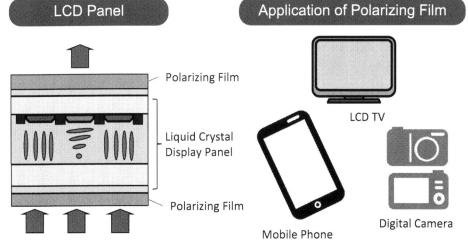

Polarizing Film

Liquid Crystal
Display Panel

Polarizing Film

LCD TV

Mobile Phone

Digital Camera

By Courtesy of Sumitomo Chemical Co., Ltd.

22

48

OA (office automation) equipment such as a copying machine have become indispensable in everyday life. Here, the principle of a copying machine is considered (see the diagram). When light is applied to negatively charged photoreceptors, those exposed to the light lose their charge. Toner attaches to the remaining positively charged electrons. Areas with a higher charge become darker. Why is charge lost when exposed to light? As the charge-generated material absorbs light and creates pairs of electrons (negative) and electron holes (positive). The charge-generating material transports the electron holes to the surface and binds them with the electrons. As a result, surface charge is negated. To accomplish this, a vital component of the photoreceptor is the charge (electron hole) transport material [22a]. Typical compounds are shown in the center left diagram. All have a tiphenylamine dimer (TPD) structure. The nitrogen atom has a lone pair of electrons, but these electrons tend to separate and become radical caions. Iodine, in the form of iodine compounds, is used for components in the compounds shown in the diagram. For example, TPD is synthesized from diiodobiphenyl and diphenylamine using the Ullman reaction (reaction formula) [22b].

Use of TPD has become popular over a wide range, from elecrophotographic materials to organic electroluminescent (EL) materials, etc.

An organic EL display is based on a four-layer (electron transport layer, luminescent layer, hole transport layer, electron injection layer) film deposited on a glass substrate and sandwiched between the two electrodes, namely, an anode and a cathode. When DC voltage is applied, light is emitted, making the display light up. A TPD derivative is used for the hole transport layer. An organic EL and a photoreceptor have different requirements. With organic EL, the higher mobility of hole transport material is the better. However, for a photoreceptor, its role as an insulator in dark conditions is more useful. Iodine is used to synthesize hole transport materials with various performance requirements [22c].

Summary Box

- Organic iodine compounds have high reactivity.

- Iodine compounds are indispensable in the synthesis of elec-tron materials.

Diiodobiphenyl

Principles of Printer

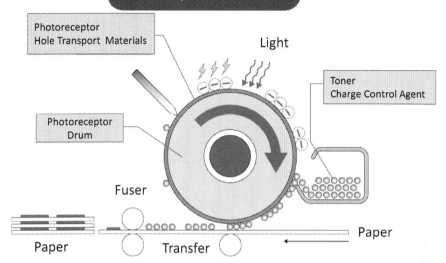

Hole Transport Materials (TPD derivatives)

NPB

TPD

Organic EL

Anode
Electron Transport Layer
Emissive Layer
Hole Transport Layer
Hole Injection Layer
Cathode (ITO)
Glass Substrate

Emission

NPB & TPD are potential hole-transport molecules owing to their great stability after oxidation (one electron removal from the lone pair of nitrogen atom).

Synthesis of Hole Transport Materials

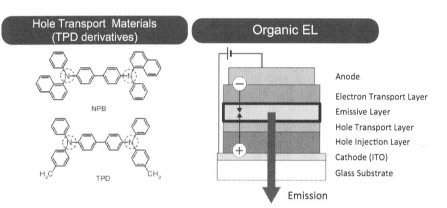

Iodine compounds are very reactive.

23

Iodine in liquid crystal displays for a clear view even at an acute angle

Wide-view film

The most widespread liquid crystal display type is the twisted nematic (TN) liquid crystal. As shown in the upper right diagram, when the voltage is applied to the TN liquid crystal, the center of the liquid crystal cell becomes charged, but the vertical edge near the glass substrate fluctuates but remains tilted at an angle. Consequently, from the side, only a fraction of the light coming in diagonally leaks from the upper part of the polarizing plate, resulting in disadvantages such as decreased contrast and loss of sharpness. To counteract these drawbacks, a wide-view (WV) film characterized by continuous tilting and aligned orientation of the disk-shaped discotic liquid crystals has been developed [23a].

A polymerizable discotic liquid crystal compound (THABB) is used for the WV film. The film is made by heating the THABB and carrying out polymerization by regulating the orientation membrane on one side only.

THABB synthesis is shown in the reaction formula. Synthesis consists of various processes, including introducing the side chain (R), creating a triphenylene ring and condensing them. Among these processes, demethylation reaction of hexamethyoxytriplenylene (HMT) is important. Only a small amount is needed, and reaction may proceed even with boron triboromide. However, this reagent is difficult to use in an industrial capacity due to reasons such as high cost, high level of corrosiveness, and susceptibility to decomposition with moisture in the air. Hydrobromic acid was also considered as an option, but reactivity was insufficient, and complete removal of the six methyl group to obtain a high-purity intermediate (HHT) was difficult. If a low-purity HHT is used to create THABB, orientation would be time consuming. At this point, hydrogen iodide (HI) was introduced. By demethylation with HI, highly pure THABB synthesis could be realized [23b,c].

By vertically sandwiching the TN crystals with a WV film, the disadvantages mentioned at the outset could be overcome, and a clear picture from any angle, vertical or horizontal, can be achieved by including liquid crystals in a WV film as shown in the diagram.

Summary Box

- Hydroiodic acid is used to manufacture WV film.

- Hydrogen iodide is an efficient demethylation agent.

Discotic Liquid Crystal Compounds (THABB)

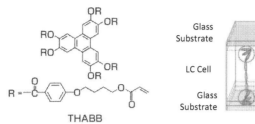

THABB

Model of Optical Compensation

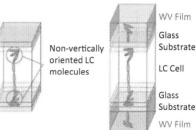

Glass Substrate
LC Cell
Glass Substrate

Non-vertically oriented LC molecules

WV Film
Glass Substrate
LC Cell
Glass Substrate
WV Film

Wide View (WV) Film & Liquid Crystal Display

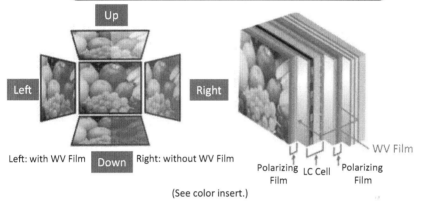

Up

Left

Right

Down

Left: with WV Film   Right: without WV Film

Polarizing Film   LC Cell   Polarizing Film

WV Film

(See color insert.)

Preparation of Intermediate of WV Film using Hydroiodic Acid

HMT

$\xrightarrow{HI}$

HHT
High Purity(> 99%) HHT was obtained.

<1%

By Courtesy of Fuji Film.

Glossary

Hydroiodic Acid: Aqueous solutions of HI are known as hydroiodicacid, a strong acid.
Hydriodicacid is a colorless liquid that turns brown on exposure to light.

51

24

Using iodine to create graphite

Graphite is comprised of multiple layers of thin plate-like carbon crystals. It shows superior heat resistance and conductivity, is lightweight, and has high tensile strength. Graphite has an increasingly broad range of use. For example, graphite molded into block form is used as battery electrodes or as the outer wall of a space shuttle. Graphite in fiber form, also known as carbon fiber, is used for F1 cars and state-of-the-art aircrafts. Currently, the mainstream method to create graphite is by high heat processing of polyacrylonitrile.

However, it is difficult to control the graphite structure at nano-level or to add a unique structure due to its insoluble and infusible nature.

Professor Kazuo Akagi et al. of Kyoto University could successfully synthesize a spiral polyacetylene (helical polyacetylene) by dispersing the polymerization catalyst on a chiral liquid crystal surface and slowly spraying acetylene gas onto it. The polyacetylene ring twists in a fixed direction and forms fibril bundles. The fibrils themselves have a hierarchical spiral structure and show high conductivity when iodine is doped.

Polyacetylene itself has low air stability, making direct carbonization difficult. However, by carbonizing the iodine-doped polyacetylene, a yield of more than 80% can be obtained, while maintaining the original form. This is possible because the hydrogen in the polyacetylene breaks away as hydrogen iodide and then builds onto the main chain, with unpaired electrons binding together forming a crosslinking action, and further developing into a crosslinking reaction at high temperatures. This method may be applied to both synthetic polymers and natural polymers [24].

A graphite sheet synthesized from cellulose can be used for the gas diffusion film for solid polymer electrolyte fuel cells.

In addition, expectations for helical graphite as an induced electromagnet are high. Furthermore, uses for circularly polarizing plates and secondary nonlinear optical materials are under consideration.

Helical graphite

52

Summary Box

- With iodine doping, graphite can be produced at low temperatures.

- The shape and form of graphite can be controlled.

Helical Graphite

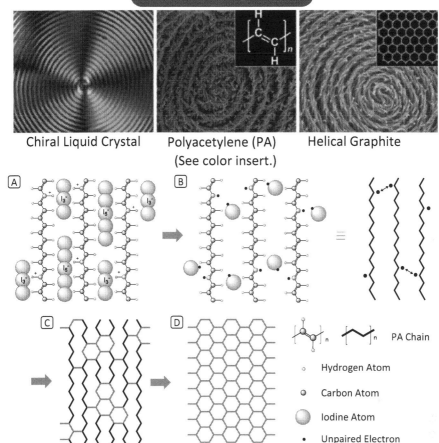

Chiral Liquid Crystal | Polyacetylene (PA) (See color insert.) | Helical Graphite

A) Structure model of iodine doped PA.
B) Removal of hydrogen as hydrogen iodide from the PA chain.
C) The cross-link formation between unpaired electrons on the backbone.
D) Helical graphite was formed by the further cross-linking reaction at higher temp.

Glossary

Fibril: Fibrils are fine fibers. Fibrils are arranged in a spiral formation at about 70 degrees to the fiber long axis.
Doping: Doping is the process intentionally adding impurities into an pure crystals to alter their properties.

53

Generally, seaweed is said to contain a high concentration of iodine, while agricultural products have a low concentration. Why is that?

Let us first examine iodine in the environment. Overall 94% of the iodine that falls to the earth in rain is iodic acid and 4% is iodide ions. This iodine accumulates in the soil and is absorbed by plants and agricultural produce. Since iodine concentration is usually higher in soil than rock, iodine is considered to be transported from the ocean to land by rain.

Iodine in the soil takes the form of iodate in most farmland, but in water immersed conditions such as rice fields, iodine is reduced to iodide ions. Although adsorption level differs according to soil type, iodine is easily adsorbed in soil. In particular, andosol soil (porous, dark volcanic soil), commonly seen in Japanese fields, adsorbs iodine most efficiently.

How about iodine concentration in agricultural produce? Migration of iodine from the soil to agricultural produce is extremely low. For example, iodine levels (mg/Agricultural Produce-kg) are as follows. Cabbage: 0.087–0.127, komatsuna (Japanese mustard spinach): 0.096, sweet potatoes 0.011, Japanese white radishes: 0.037–0.123, eggplants: 0.041, carrots 0.076, tomatoes: 0.041, onions: 0.12.

A study on iodine-enriched agricultural produce was carried out where the iodine concentration in agricultural produce was increased. To do so, iodine was added to the fertilizer or when watering.

However, it is also important to consider the concentration since high concentrations may inhibit growth in plants. Up to now, iodine-enriched lettuce and tomatoes were successfully obtained through nutriculture without hindering growth.

In addition, iodine enrichment through organic fertilizers (seaweed compost or iodine poultry manure) is another method. For example, iodine concentration became 1.1–4.9 mg/kg with the use of seaweed compost and 0.8–1.4 mg/kg with iodine poultry manure.

In regions not accustomed to eating marine products, iodine can be supplied to the diet by iodine-enriched products.

Iodine enriched tomatoes

4 Using iodine for analysis

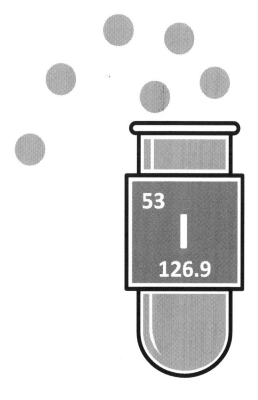

25

Chemical reactions learned in science class

Iodine-starch reaction

Starch is a glucose polymer (D-glucose) comprises of approximately 20% liner polymer amylose and 80% amylopectin polymers with many branches. In the iodine-starch reaction, amylose molecules in the starch form a helical structure in the aqueous solution and turn bluish purple-reddish brown when iodine molecules are present. When the colored solution is heated, the iodine molecules are released from the helical structure, and the solution becomes colorless. However, when cooled, the helical structure is restored as iodine molecules return, and the solution regains its color. The iodine-starch reaction is well used in the educational field, and is always quoted in elementary, junior high, and high school science, chemistry and biology textbooks, as well as in food science textbooks. The iodine-starch reaction is used to verify the existence of starch from photosynthesis in plants leaves and to track the digestion and decomposition of starch by digestive enzymes. In addition, iodine plays a vital role in analytical chemistry. For example, the iodine titration is a typical titration method in volumetric analysis along with neutralization titration, oxidization–reduction titration, and precipitation titration [25].

The iodine titration method is described as follows: By adding an oxiddizing agent such as copper sulfate into a potassium iodide solution, iodide ions (I^-) are easily oxidized to iodine (I_2). The quantity of free iodine can be determined using a reducing agent of a known concentration, and in turn, the quantity of the oxidant can be determined indirectly. In general, standard sodium thiosulfate ($Na_2S_2O_3$) solution is used for the titration of free iodine (see reaction formula). The end point of this reaction can also be determined by the loss of iodine color. If a dark blue color is generated by adding the starch solution and the end point determined when the color disappears, the concentration can be clearly verified. For example, this reaction is used for the quantification of formalin, copper, phenol, etc.

Summary Box

- Starch takes in iodine in a helical structure and emits color.

- Starch is used as an indicator in quantitative analysis.

Components of Starch

Amylose

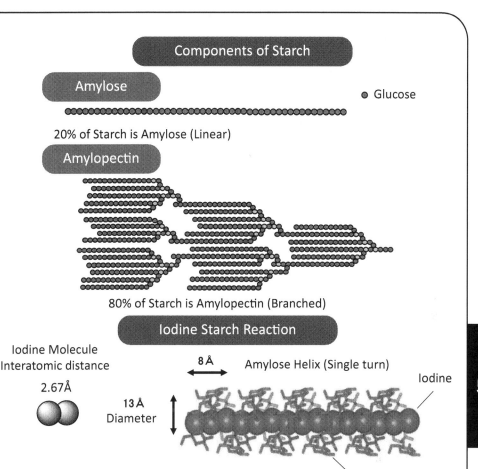

● Glucose

20% of Starch is Amylose (Linear)

Amylopectin

80% of Starch is Amylopectin (Branched)

Iodine Starch Reaction

Iodine Molecule
Interatomic distance

2.67Å

8Å   Amylose Helix (Single turn)

13Å
Diameter

Iodine

Amylose

3.2Å   Polyiodide
Interatomic distance

Polyiodide slip into the amylose helix.
Amylose iodine complex : bluepurple –redpurple (depending on
the molecular weight of amylose and the length of polyiodide chain)

Titration Reaction of Iodine

$$4KI + 2CuSO_4 \rightarrow CuI_2 + 2K_2SO_4 + I_2$$

$$I_2 + 2Na_2S_2O_3 \rightarrow 2NaI + Na_2S_4O_6$$

The liberated iodine (in first step) is titrated with standard solution of
sodium thiosulfate. Starch is used as indicator to increase sensitivity .

26

Measuring iodine concentration

Various analysis methods of iodine

Iodine exists in various forms in the natural environment in the atmosphere, ocean, and soil, as well as in other environments such as in minerals, food, and chemical products, and plays a vital role in the respective environments. How are iodine and its compounds analyzed? In this section, the analysis methods of iodine are considered. The concentration of iodine in the environment is low. Therefore, a special filter is used to collect iodine, and extraction from the filter is carried out using a warm sodium hydroxide/sodium sulfite solution with radioactive ^{129}I added as an internal standard. By adding silver sulfate, iodine is precipitated as silver iodine (AgI). Iodine deposits are dissolved in ammonia water, and analyzed using isotope dilution mass spectrometry.

For iodine in water, seawater, or wastewater, chloric acid, bromine, and pergmanganic acid are used to convert iodide in the sample to iodate. Next, potassium iodide (KI) is added, and any generated iodine is titrated by sodium thiosulfate (see Section 25). In another method, the sample is first acidified with hydrochloric acid, then oxidized with hydrogen peroxide or potassium permanganate. After removing the excess oxidant, the generated iodine is measured with a spectrophotometer.

Iodine-added salt may be measured using an ion chromatography (see Section 27). However, there is a nondestructive measurement method using fluorescent X-ray spectroscopy. Low-concentration iodine can be measured using a recent energy dispersive device (upper diagram).

For iodine in plants, pretreatment of the sample by the microwave digestion method using acetic acid and hydrogen peroxide is carried out, and after reduction treatment with sodium thiosulfate or ascorbic acid, iodate is converted to iodide. The iodide is measured using an inductively coupled plasma mass spectrometer (ICP-MS) (lower diagram) [26a,b].

There are still other easy-to-use colorimetric analysis methods. An iodine concentration measurement device which uses N, N-diethyl-p-phenylamine (DPD) as an indicator is available from Hanna Instruments (middle diagram) [26c].

Summary Box

- Iodine measurement methods: Mass spectrometry, titration method, fluorescent X-ray spectroscopy, colorimetric method.

X-ray Fluorescence Analysis

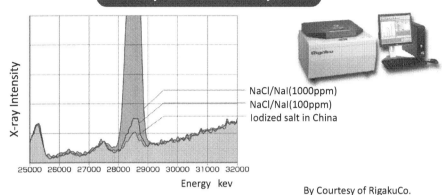

NaCl/NaI(1000ppm)
NaCl/NaI(100ppm)
Iodized salt in China

X-ray Intensity

25000 26000 27000 28000 29000 30000 31000 32000
Energy kev

By Courtesy of RigakuCo.

	Quantitative value	Theoretical value
Iodized salt	43	20–50
NaI 100 ppm /NaCl	80	85
LiI 100 ppm /NaCl	95	95

Iodine meter (Hanna)

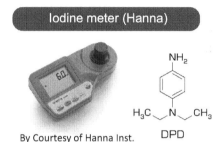

By Courtesy of Hanna Inst.
(See color insert.)

DPD

Principles of ICP-MS Instrument

Pump

Quadrupole   Lens System   Corn   High-frequency coil   Torch

③ Detector   ①

Plasma   Chamber   Argon Gas   Sample Solution

②

| Detecting System | Mass Spectrometer | Lens System | Interface | Ionization System | Sample Injection System |

① Spray a liquid sample in the high temperature inductively coupled plasma (ICP) .
② Introduce a heated decomposed ionized atoms in the plasma to a mass spectrometer (MS).
③ Detect molecular mass (m/z) and conduct quantitative analysis.

27

How to measure iodide ion concentration

Ion chromatography

Ion chromatography is a type of liquid chromatography which can qualify and quantify anions such as chloride ions, fluoride ions, and sulfate ions and cations such as sodium ions and ammonia ions with high sensitivity. Ion chromatography uses ion-exchange resin as the stationary phase. By utilizing the difference in the time each substance remains on the ion-exchange resin according to the strength of the charge, substances within a sample can be separated. In the suppressor between the column and the detector, ions which were originally included in the mobile phase are removed and placed on the detector.

There are various kinds of detectors such as electrochemical detectors, UV/visible absorbance detectors, and electric conductivity detectors. In addition, to analyze iodide ions and iodate ions, an anion analysis column can be used.

Electrochemical detector: Iodide ions are oxidized on a silver electrode when voltage is applied. The oxidation current is measured with an electrochemical detector. Ions such as chloride ions do not readily oxidize or reduce on the electrode and are hardly detected. Therefore, this detector is best suited to measure trace amounts of iodine ions in seawater.

UV/visible absorbance detector: The iodide ion has absorbance in the ultraviolet region and can be detected at a wavelength of 226 nm. This detection method is effective when measuring iodide ions in samples with a large amount of chloride ions and sulfate ions which do not have absorbance in the ultraviolet region.

Electric conductivity detector: In comparison to other detection methods, the peak intensity of iodide ions by this method is low. However, it has the advantage that other inorganic ions can be simultaneously detected.

As shown above, any of the above detectors can be used for iodide ion analysis. Ion chromatography is an indispensable method to analyze iodide ions in brine, the raw material of iodine [27a,b].

Summary Box

• Ion chromatography is used to separate anions.

• There are many detection methods for iodide ions.

Principles of Ion Chromatography

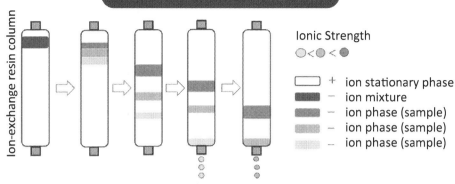

Ionic Strength

○ < ◉ < ●

☐ + ion stationary phase
▬ − ion mixture
▬ − ion phase (sample)
▬ − ion phase (sample)
▬ − ion phase (sample)

Ion Chromatography Flow Diagram

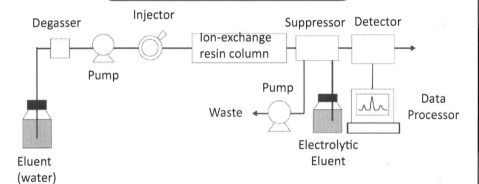

Degasser   Injector   Suppressor   Detector

Pump   Ion-exchange
                      resin column

                              Pump
                  Waste

                              Electrolytic
Eluent   Eluent
(water)

Data
Processor

Ion Chromatography Graph

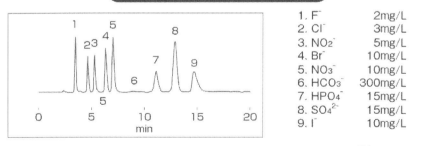

1. F⁻	2mg/L
2. Cl⁻	3mg/L
3. NO₂⁻	5mg/L
4. Br⁻	10mg/L
5. NO₃⁻	10mg/L
6. HCO₃⁻	300mg/L
7. HPO₄⁻	15mg/L
8. SO₄²⁻	15mg/L
9. I⁻	10mg/L

By Courtesy of Shodex™.

28

62

The properties of fats and oils can be clarified with iodine

Iodine value

The iodine value is an indicator used to evaluate the properties of oils, fats, and biodiesel fuel. The carbon-carbon double bond within an organic compound has high reactivity, and its deterioration may cause a fire and oxidation by air may cause heat generation.

The iodine value is mainly used to compare the carbon-carbon double bond contained in complex mixtures of natural origin, such as animal and vegetable oils. The iodine value is expressed by converting the halogen quantity into the number of grams that reacts with 100 g of substance.

The iodine value can be obtained by adding an excessive iodine to the sample and allowing it to completely react, and then determining the quantity of the remaining iodine by oxidation–reduction titration. In reality however, the reactivity of iodine alone is insufficient. Therefore, the Wijs method with iodine monochloride is used as a reagent or the Hanus method with iodine bromide is adopted. The standard test method for the iodine value is specified under industrial standards [28a].

As shown in the table, vegetable fats and oils, based on the iodine value, are classified as drying oil (iodine value of 130 or higher), semidrying oil (iodine value of 100–130), or nondrying oil (iodine value of 100 or less), according to their drying property. Fats and oils that are easily oxidized and when left alone, result in a reaction as shown in the lower diagram, become resinous, and form a coating on the surface. This indicates a high drying property. On the other hand, fats and oils that do not easily oxidize maintain their liquid form and are referred to as having a low drying property.

As show above, there is a correlation between the iodine value and drying property, and nondrying oils are considered to be fats and oils with an extremely low drying property. In cosmetics, nondrying oil from vegetable fats and oils is most widely used.

Furthermore, the iodine value is also used as an indicator for deterioration in the properties of fats and oils. As deterioration progresses, the value decreases [28b].

Summary Box

- The iodine value is used to express the content of the double bond in animal and vegetable fats and oils.

- Fats and oils are classified according to their iodine value.

Iodine Value

$$-[CH_2-CH=CH-CH_2]_n + nI_2 \longrightarrow -[CH_2-\underset{|}{CH}-\underset{|}{CH}-CH_2]_n$$

M:Molecular Weight of Oil
Molecular Weight of Iodine: 254
Number of Double Bonds: n
Iodine Value : y

$$y = \frac{25,400n}{M}$$

Iodine value is the number of grams of iodine consumed by 100g of oil.
A higher iodine value indicates a higher degree of unsaturation.

Vegetable Oil Classification Iodine Value

Vegetable oil	Iodine Value
Nondrying oil	<100
Coconut oil	15~25
Palm oil	30~40
Sunflower oil	80~90
Semidrying oil	100~130
Rapeseed oil	100~120
Soybeen oil	115~140
Drying oil	>130
Tung oil	140~180
Linseed oil	160~190

Oxidation of Oil

The oil is oxidized in the air and becomes resin.
①Autoxidation of Diene (unsaturated oil) produces hydroperoxide.
②Hydroperoxide cross links with another unsaturated side chain.

29

Moisture analysis and iodine

64

The Karl Fisher method

The Karl Fischer (KF) method and the loss-on-drying method are the main methods to measure moisture in a substance. Of these, the KF method is widely established as the most reliable chemical method and is used not only as the moisture test for various chemical products, but also for various industrial products. It has been adopted as both the international standard and the domestic standard (Japan Industrial Standard [JIS], Japan Agricultural Standard [JAS], and Japan Pharmacopoeia [JP]).

This method was invented in 1935 by Karl Fischer in Germany [29a]. The reagent used in this method is a mixture of iodine, sulfur dioxide, pyridine, and methanol, and is collectively known as the "KF reagent," named after the inventor. This KF reagent can selectively and quantitatively react with water, with reaction occurring with 1 mole of iodine and 1 mole of water. Basically, this reaction formula is still recognized today, and the reaction formula (1) in which pyridine is expressed as the base (RN) has become commonplace.

Many automatic analysis devices have been developed based on this principle and have gained popularity. As shown in the diagram, there are two titration methods, the volumetric method and the coulometric method.

Compared to the loss-on-drying method, the distillation method, the infrared absorption method, the electrolytic method, and other electric methods, the KF method has the following superior characteristics: (1) A calibration curve is not required and the absolute moisture value can be obtained (with the coulometric method). Minute moisture quantities at the ppm scale can be accurately determined. (2) Measurement can be quickly carried out. (3) Only a small amount of reagent (few mg to few g) is required. (4) Can be applied to liquids, solids, or gas reagents. (5) Can be applied to unstable substances that change when heat is applied.

Moisture measurement devices that apply the KF method using iodine are widely utilized in plant laboratories and research institutions such as chemical, pharmaceutical, and food factories [29b].

Summary Box

- Iodine is used in moisture measurements which are indispensable in factories and research institutions.

- Iodine reacts with moisture.

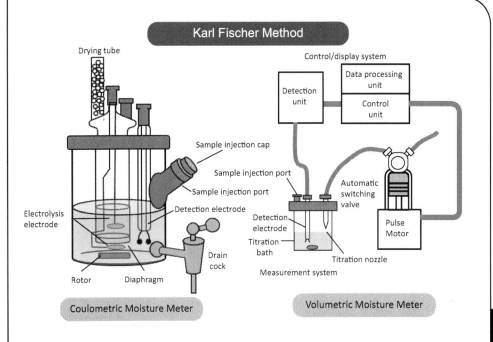

Karl Fischer Method

Coulometric Moisture Meter

Volumetric Moisture Meter

Volumetric Titration

$$H_2O + I_2 + SO_2 + CH_3OH + 3RN \rightarrow 2RN \cdot HI + RN \cdot HSO_4CH_3 \quad (1)$$

Water (mg) = Titration volume (mL) x Karl Fischer Titer (mgH$_2$O/mL)
Karl Fisher Titer: the number of mg of water (H$_2$O) in 1 mL of Water — Methanol
Standard solution.

Coulometric Titration

$$2I^- - 2e \longrightarrow I_2 \quad (2)$$

Water 1 mg=10.71 coulombs

By Courtesy of Mitsubishi Chemical Analytech Co. Ltd.
(See color insert.)

30

Using iodide to measure radiation

Scintillators

66

On March 11, 2011, the Tohoku region was hit by the Great East Japan Earthquake and subsequent tsunami. In its wake, a serious accident involving the release of radioactive materials occurred at the Tokyo Electric Power Company's Fukushima Daiichi Nuclear Power Plant. Immediately after the accident and for a few days after, radiation peaked in the Kanto Region including Fukushima Prefecture, and radioactive substances as iodine 131, cesium 134, and cesium 137 originating from the accident were detected in the atmosphere and soil, affecting food and drinking water as well. These radioactive substances release β and γ rays.

There are various devices to measure radiation according to the type and intensity of radiation, such as α, β, and γ rays. Among these devices, the scintillator-type survey meter measures γ rays.

The scintillator-type survey meter reacts with γ and X-rays, and measures the radiation energy and dose using a substance which emits a weak light (scintillator) and a photomultiplier tube which amplifies the light.

The measuring efficiency per unit volume of the γ ray survey meter depends on the electron density in the survey meter. In particular, iodides such as sodium iodide (NaI) and cesium iodide (CsI) contain elements with high atomic numbers, resulting in a high count efficiency.

In addition, particles with small amounts of thallium added to both NaI and CsI to increase the light yield, are used. However, one disadvantage of both of these is their deterioration when exposed to water and high humidity.

This device can measure radiation from 0.1 μSv/hour to several tens of μSv/hour. Since naturally occurring potassium 40 and cobalt 60 also emit γ rays, such natural radiation will also be included in the measurement [30a,b,c] if they are not filtered out.

Summary Box

- Iodides play a central role in γ ray scintillators.

- Iodides are susceptible to moisture.

Principles of Scintillation

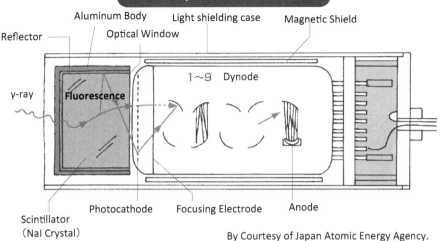

By Courtesy of Japan Atomic Energy Agency.

Energy photon (Gamma) hits a scintillating crystal and triggers the release of low-energy photons (fluorescence) which are then converted into photoelectrons and multiplied in the photomultiplier.

Single crystal for Scintillator

Single crystal of CsI
By Courtesy of Leading Edge Algorithms Co., Ltd.
(See color insert.)

Scintillation Survey Meter

By Courtesy of Hitachi Ltd.

Glossary

Radioactive substances : After nuclear power plant accident in Fukushima , Radioactive substances (^{131}I, ^{134}Cs, and ^{137}Cs) were detected . Nearby residents were forced to evacuate.

31

68

As incident light and molecules interact, light is scattered (Raman effect) and the vibration frequency of the incident light fluctuates. The Raman spectrum analysis uses this phenomenon to obtain information on the molecular and crystal structures of a substance. This effect was discovered by Sir Chandrasekhara Venkata Raman (1880–1970, India) who received the Nobel Prize in Physics in 1930.

If properly done, the Raman spectrum analysis can make a quantitative assessment from the spectral intensity. Raman spectrum analysis is extremely effective as a method to analyze iodine compounds. Graph (1) shows an example of the quantitative assessment of iodine in acetonitrile. From the Raman spectrum of acetonitrile solutions with different iodine concentrations, an iodine peak of around 200 cm^{-1} is detected, theoretically indicating the existence of iodine in the I_2 state in the solution. Furthermore, as indicated in the diagram, from the iodine concentration in the acetonitrile and the peak area ratio (iodine/acetonitrile), iodine concentration can be seen to have a good linear relationship. From the above, within the concentration shown above, a quantitative assessment of iodine (I_2) in acetonitrile can be made.

Photo (2) shows an imaging example of the TiO_2 electrode film surface of a dye-sensitized solar cell. The existence of iodine (I_2), an electrolytic substance, is verified using the Raman spectrum.

Furthermore, iodine within diagram (3) polarizing film contains two types of isomers, namely I_3^- and I_5^-. By adjusting the concentration of these two types of isomers, the color tone of the polarizing film changes. Using the results of the 2D (two-dimensional) cross-section mapping using Raman spectroscopy, the component ratio of the iodine isomers was shown to differ near the center and edge of the iodine layer. In addition, near the interface of the film layer, a layer with a small quantity of I_5^- was also found [31a,b,c].

Summary Box

- Raman spectrum analysis is used to measure iodine concentration in solutions and crystals.

- Raman received a Nobel Prize in Physics.

Raman Spectroscopy

$E = h\nu$
E: Energy
h: Planck's Constant
v: Frequency

Incident light (ν_0)

Molecular vibration
Energy ν_1

$\nu_0 + \nu_1$
Raman scattered light
(Anti-Stokes light)

ν_0
Rayleigh scattering light

$\nu_0 - \nu_1$
Raman scattered light
(Stokes light)

Raman Spectroscopy of Iodine

(1)

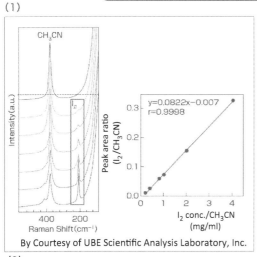

$y = 0.0822x - 0.007$
$r = 0.9998$

CH$_3$CN

Intensity (a.u.)

Peak area ratio (I$_2$/CH$_3$CN)

Raman Shift (cm^{-1})

I$_2$ conc./CH$_3$CN (mg/ml)

By Courtesy of UBE Scientific Analysis Laboratory, Inc.

(2)

Raman mapping images of TiO$_2$ film electrode surface

By Courtesy of Renishaw K.K.
(See color insert.)

69

(3)

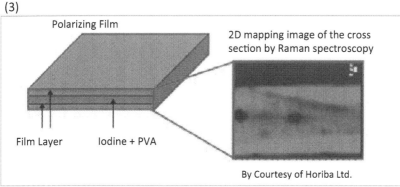

Polarizing Film

2D mapping image of the cross section by Raman spectroscopy

Film Layer   Iodine + PVA

By Courtesy of Horiba Ltd.

32

70

Analyzing solid samples with low iodine content, such as soil, without compromising the sample, is difficult. A method of analysis which can do this is the X-ray absorption fine structure (XAFS) analysis. In Japan, there are several huge synchrotron radiation facilities for XAFS measurement. Among them, Spring 8 in Hyogo Prefecture [32a] and Photon Factory (PF) in Tsukuba City [32b] are internationally famous.

When the direction of electrons moving at speeds close to the speed of light are changed magnetically, synchrotron radiation is generated. The higher the electron energy, the better the directivity and the brighter the light becomes. In addition, the greater the direction of movement is changed, and more synchrotron radiation with a short wavelength such as an X-ray is created. As an accelerator, Spring 8 has a synchrotron with a circumference of 396 m and a storage ring with a circumference of 1,436 m.

Specific applications of iodine analysis are described below. To examine iodine concentration and its form in soil, the common practice is to extract iodine from the soil and analyze the iodine dissolved in a solution. However, the chemical procedure of extraction from soil may change the form of the iodine.

However, in X-ray near-edge structure (XANES) analysis, light that is approximately 10 billion times brighter than normal light source X-rays, and X-ray energy that excites only iodine atoms (iodine K absorption edge: 33.2 keV) can be used to irradiate a soil sample containing iodine. As a result, only the information related to iodine form is shown (diagram) in the X-ray absorption spectrum.

In the environment, four forms of iodine (as shown in the diagram) potentially exist: iodate ions (IO_3^- pentavalent), iodide ions (I^- monovalent), molecular iodine (I_2 0-valent), and organic iodine (monovalent). However, differences in their spectrum show that in the soil, iodine is mainly found in the form of iodate ions [32c].

Synchrotron radiation
X-ray absorption
spectrometry

Summary Box

- Synchrotron analysis is used to analyze iodine in the soil.

- Various forms of iodine can be measured.

SPring-8 (Large Synchrotron Radiation Facility)

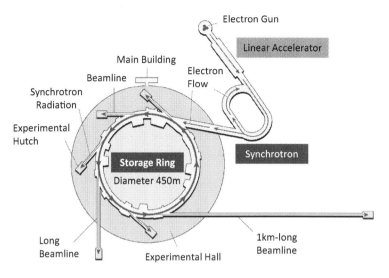

Electron Gun

Linear Accelerator

Main Building

Beamline

Electron Flow

Synchrotron Radiation

Experimental Hutch

Storage Ring
Diameter 450m

Synchrotron

Long Beamline

1km-long Beamline

Experimental Hall

By Courtesy of Riken Japan.

XANES Spectroscopy of Iodine Compounds

Absorbance

Potassium iodate
Potassium iodide
Molecular iodine
Organic state iodine

33.0   33.1   33.2   33.3
X-ray Energy (keV)

Iodine Analysis in Rice Field

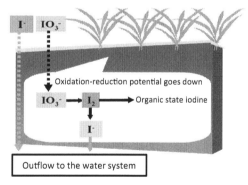

I^- IO_3^-

Oxidation-reduction potential goes down

IO_3^- ⟶ I_2 ⟶ Organic state iodine

I^-

Outflow to the water system

By Courtesy of NIAES.

Quinoform is an iodine-containing drug that was widely sold in Japan as an intestinal remedy, but was found to cause SMON (subacute myelo-optico-neuropathy) disease, and its production was discontinued in 1970. It is currently available in some countries as specific medicine for amoebic dysentery. This drug is a potent copper/zinc chelating agent and removes metal from intercellular metalloenzymes. However, one side effect is the loss of vitamin B_{12}, which is theorized to trigger the onset of SMON disease.

However, this drug is currently the focus of attention as a remedy for Alzheimer disease. In Alzheimer disease, β-amyloid protein (Aβ) is deposited in the brain, and copper ion and zinc ion play a role in this Aβ deposition (insolubilization). In model mice that were administered Quinoform, this Aβ deposition dissolved. Among metal chelating agents, Quinoform has the highest copper and zinc removal effect. From this, a research group at the Pathology Department of the University of Melbourne in Australia headed by Cleve W. Ritchie is conducting a placebo-controlled trial study on the effect of Quinoform on Alzheimer disease. In a clinical test on moderate to advanced stage Alzheimer patients who received Donepezil hydrochloride (product name in Japan: Aricept) treatment for 6 months, Quinoform administration significantly suppressed the decrease of cognitive function in advance stage patients. Furthermore, during the test period, neither vitamin B_{12} or folic acid deficiency, nor the onset of SMON disease, was observed.

From the above findings, the research group concluded that the concept of "Alzheimer disease chelation therapy" using Quinoform or similar compounds is promising.

Quinoform

Monument of Iodine Discovery
Section 3

Plaque on the Courtois' home
Section 3

Courtois' home in Dijon, France
where Bernard Courtois was born
Section 3

Remains of a seaweed kiln
on the coast of Brittany
Section 3

Flake Iodine
Section 7

Iodine Mine & Iodine Plant in Chile
Section 8

Prilled Iodine
Section 7

Caliche Ore
Section 8

Gargle Mouth wash
Section 11

Amycel Iodine Mask
Section 13

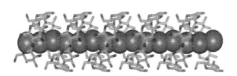

Amylose Iodine Complex
Section 13

Cyclodextrin Inclusion
Complexes
Section 14

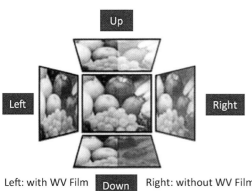

Up

Left

Right

Left: with WV Film Down Right: without WV Film

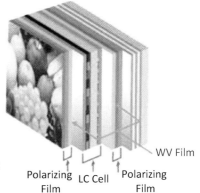

WV Film

Polarizing LC Cell Polarizing
Film Film

Wide View (WV) Film & Liquid Crystal Display
Section 23

Chiral Liquid Crystal

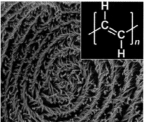
Polyacetylene (PA)

Helical Graphite

Section 24

Iodine Meter

Section 26

Karl Fischer
Moisture Meter

Section 29

Single Crystal of CsI
Section 30

Raman mapping images of
TiO$_2$ film electrode surface
Section 31

Acetic acid production facility
Section 33

Ionic Agent
○ Iodine
○ Side Chain

Nonionic Agent
(Monomer Type)

Nonionic Agent
(Dimer Type)

X-ray contrast agent
Section 47

Potassium iodide Pills
Section 48

Ioflupan (^{123}I) Injection
Section 49

Dosage Form of Cadexomer Iodine
Section 51

ISAN system
Section 57

DSSC Cell
Section 59

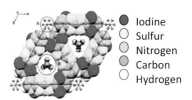

- Iodine
- Sulfur
- Nitrogen
- Carbon
- Hydrogen

XB & Superconductors
Section 64

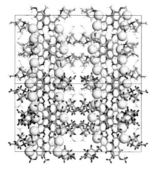

Herapathite
Column 2

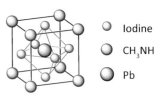

- Iodine
- CH$_3$NH
- Pb

Perovskeit
Column 8

Innovative industrial technology starts with iodine

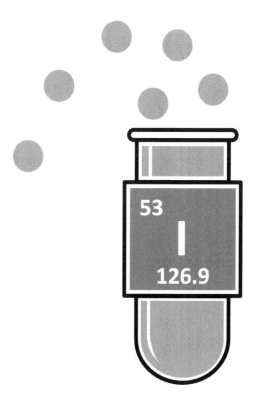

33

Acetic acid is an important industrial material with an annual production of 6.5 megatons. Polyvinyl acetate, a typical adhesive agent, is produced by the polymerization of vinyl acetate monomer, which is synthesized from polyvinyl acetate and ethylene. In addition, ethyl acetate which is used as a solvent for paint and printing ink, and ester acetates such as butyl acetate and propyl acetate, are produced from acetic acid and various types of alcohol.

Edible acetic acid and vinegar are produced from ethanol by a fermentation process. As various methods to produce industrial acetic acid were developed over the years, the raw materials have changed significantly.

Until the early twentieth century, pyroligneous acid (wood vinegar) obtained from the dry distillation liquid of wood was used as a raw material for industrial acetic acid. Since then, acetylene obtained from coal and ethylene were progressively used as the petrochemical industry expanded.

In 1960, an epoch-making approach to make synthetic acetic acid was developed using methanol and carbon monoxide. However, harsh conditions, namely 700 atm and 300°C, were required. To resolve these conditions, the Monsanto Company developed a Monsanto catalyst (see the diagram) which combined iodine and rhodium in 1966, where moderate reaction conditions of 30–60 atm and 150–200°C could be used.

Later, BP Chemicals Ltd. developed a Cativa catalyst (see the diagram) combining iodine and iridium. Under the Cativa method, acetic acid is produced by combining methanol and carbon monoxide. Methyl iodide (CH_3I), acetyl chloride (CH_3COCl), hydrogen iodide (HI), and iridium complex are formed as intermediates and have high reactivity, and efficiently produce acetic acid. The Cativa method, in comparison to the Monsanto method, requires less water in the reaction mixture, produces less acid, and uses simplified production facilities. Presently, most industrial acetic acid is produced using this method [33a,b].

Iodine used to synthesize acetic acid

Iodine iridium complex

74

Summary Box

- Catalyzed acetic acid synthesis requires hydroiodic acid.

- Change of metal improved reaction yield.

Catalyst for Acetic Acid Production

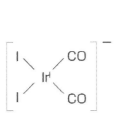

Catalyst for Monsanto Process

Catalyst for Cativa Process

Acetic acid production facility
(1: reactor, 2: distillation column)
Photo provided BP Chemicals Inc.
(See color insert.)

The catalytic cycle for the Cativa process

CH₃OH ①→ CH₃-I

HI   H₂O

②

③

④

⑤

⑥

H₃C-C(=O)-OH   H₃C-C(=O)-I

Iodine Catalyst

Acetic acid is efficiently produced according to the reactions ①→⑥.

34

Iodine: Indispensable for the synthesis of water repellents

76

Telogen

Many items we carry or wear, such as bags, shoes, hats, umbrellas, and clothes, as well items such as sofas and carpets, are treated with a fluoropolymer agent as a water and oil repellent, in order to provide products that are water, oil, and dirt resistant. In addition, this treatment deters organisms including microbes from attaching themselves to the fluoropolymer surface, and even if they do, they cannot eat and digest the fluoropolymer. As a result, fluoropolymer exhibits an antifouling effect.

The brilliant balance in attraction between carbon and fluoropolymer electrons which constitute the fluoropolymer, $-CF_2-$, makes the fluoropolymer virtually nonpolar (surface free energy—10 mJ/m^2), minimizing interaction with substances which come in contact with the surface (aqueous liquids such water, coffee, and juice and oil-based liquids such as oil, gasoline, etc.).

In order for fluoropolymers to manifest such superior properties, the role of iodine in their production process is very important. A typical production process of fluororesin is known as telomerization (see the diagram). (1) iodine (I_2) and iodine pentafluoride (IF_5) react in pentafluoroethylene to create pentafluoroethyl iodide (C_2F_5I), known as telogen. (2) the elongation reaction of tetrafluoroethylene is used to create a fluorocarbon chain from the telogen. (3) Ethylene is added to form perfluoroalkyl iodide, which modifies iodine into a hydroxyl, and fluorotelomer alcohol is obtained. (4) Ethyl acrylate is composed from perfluoroalkylethyl alcohol and acrylic acid, and by reacting this with other comonomers (acrylate, metacrylate, vinyl chloride, vinylidene chloride), water and oil repellent fluoropolymers are obtained [34]. In addition to C_2F_5I, heptafluoropropyl iodide(($CF_3)_2CFI$) is also used as a telogen.

In the sequence of reactions, iodine does not remain in the polymers and can thus be recycled.

Summary Box

- Iodide is the starting point for water and oil repellent fluoropolymers.

- Iodine does not remain in the fluoropolymers and is recycled.

The catalytic cycle for the Cativa process

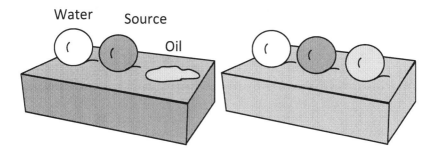

Water   Source

Oil

Non-fluorine-based water repellent
Water, the source repel, but the oil is not repelled

Fluorine-based water repellent
Repel water, source, also with oil

Synthesis of Fluorine Resin & Iodine

Synthesis of Telogen

$$2IF_5 + 10C_2F_4 + 4I_2 \rightarrow 10\boxed{C_2F_5I} \qquad (1)$$

Telogen

Telomerization reaction

$$C_2F_5I + n\,CF_2{=}CF_2 \rightarrow C_2F_5(CF_2CF_2)_nI \equiv \boxed{R_fI} \quad (2)$$

Telomer

Transformation of Telomer and Polymerization

$$R_fI \rightarrow R_fCH_2CH_2I \rightarrow R_fCH_2CH_2OH \qquad (3)$$

$$\rightarrow CH_2{=}CHCO_2C_2H_4R_f \rightarrow \text{Fluorine resin} \qquad (4)$$

Iodine compounds are highly reactive as the carbon-iodine bond is weak.
Iodine is recovered and recycled without remaining in the fluorine resin.

Glossary

Telomer: Fluorotelomers are fluorocarbon-based oligomers
Telogen: Telogens (perfluoroalkyl iodides) are starting materials for the preparation of telomers.

35

Iodine: Behind global inventions

Conductive polymers

Every day, we benefit from various synthetic polymers. For example, we use PET (polyethylene terephthalate) plastic bottles, polyethylene shopping bags, and PVC (polyvinyl chloride) water pipes. Under normal conditions, these polymers are insulators and do not conduct electricity. However, various conductive polymers are now being developed.

Dr. Hideki Shirakawa is the first person in the world to invent conductive polymers, for which he received a Nobel Prize for Chemistry in 2000. Iodine plays an important role in this invention. At first glance, polyethylene and polyacetylene appear to be similar (see the diagram). However, the main polyethylene chain is made up only of a strong single bond (σ bond), whereas the main polyacethylene chain has a conjugated structure, composed of alternating single (σ bond) and double (π bond) bonds. This conjugated structure affects whether the polymer is conductive or not. The electrons involved in this bond are called σ electrons and π electrons, respectively.

π electrons, compared to σ electrons, are widely spaced, but unlike free metal electrons, cannot move freely. However, when a reagent (acceptor) which easily accepts the electrons is added, π electrons become easily detached. If an acceptor such as iodine is added to polyacetylene, the iodine extracts the π electron from the main chain, creating a hole with a positive charge.

To illustrate, this is like a few people getting off of a crowded commuter train where there is no room to move around. People quickly fill the space as others get off, and there is room to move around. Likewise, when voltage is applied, an electronic relay becomes possible as the holes move one after another, causing the π electrons to move, thus conducting electricity.

Based on this concept, conductive polymers such as polythiophene and polypyrrole were developed and are used in various electric and electronic products [35].

Summary Box

- A conjugated polymer structure is the key to conductivity.

- To achieve conductivity iodine is added.

Iodine Doped Polyacetylene and Electro-conductive Polymers

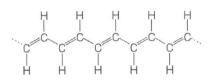

Polyethylene

Dr. Hideki Shirakawa

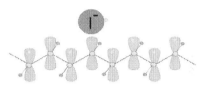

Polyacetylene

p-orbital of iodine doped
poly-acetylene (image)

Electrical Conductivity of Polyacetylene (Image)

A thin film of polyacetylene

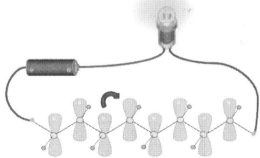

Electric Current

By Courtesy of Sigma-Aldrich Japan.

36

80

Polyamide fibers are generally known as nylon, and on account of their superior toughness, adhesiveness, and fatigue resistance, they are widely used for various industrial materials such as rubber reinforcement cords for tires, conveyor belts, transmission belts and rubber hoses, safety belts, tents, plaited cords, sewing thread, and airbags. In 2011, global production of polyamide fibers reached 6.8 million tons [36a]. Polyamides used as industrial materials are exposed to harsh environments during the processing stages of production such as heat, oxygen, and light and during use as automobile products (particularly nylon cords for tires and airbag cords). In order to prevent deterioration in these environments, various stabilizers are added.

Stabilizers can be roughly divided into copper halides, phenols, and amines. These have various characteristics as shown in the table. Copper halides, particularly copper iodides (CuI_2), are better under high-temperature conditions than other organic stabilizers and are widely used. Iodine compounds are most commonly used in the form of potassium iodides and sodium iodides [36b].

There are many studies on the mechanism of copper iodides as a stabilizer, and the mechanism most commonly held is shown in the diagram. When polyamides are exposed to high temperatures, the methylene group next to the amide group is attacked by oxygen and changes into a highly reactive structure known as a peroxy radical. As a result, a chain reaction occurs causing the amide bond to decompose. Copper iodides are thought to interact with the nitrogen and oxygen atoms of the amide group at this time, preventing radicals from occurring. As a result, decreased tensile strength is avoided [36c,d].

By further adding copper ions or iodine, antibacterial properties can be simultaneously added to the resin. In addition, iodine is also used as a stabilizer for paper bulking agents, ink resins, tall oil fatty acids, tall oil rosin, etc.

Summary Box

- Copper iodides stabilize polyamides.

- Copper iodides are superior in high-temperature environments compared with other organic stabilizers.

Application of Polyamide

Tire Cord

Airbag Wire

Mechanism of the Degradation of Polyamides

$$\sim CH_2CH_2NH_2\overset{\overset{\displaystyle O}{\|}}{C}\ CH_2CH_2 \sim$$

$$\downarrow O_2/h\nu$$

$$\Delta/O_2 \qquad \downarrow \qquad \Delta/H_2O$$

$$\sim CH_2\overset{\overset{\displaystyle O}{\|}}{C}OH \quad + \quad NH_2\overset{\overset{\displaystyle O}{\|}}{C}\ CH_2CH_2 \sim$$

"Degradation due to the effects of oxygen, heat and light"

Polyamide Polymer

$$\sim\!\!\left[CO(CH_2)_4CONH(CH_2)_6NH \right]_n\!\!\sim$$

Nylon 66

$$\sim\!\!\left[CO(CH_2)_5NH_2 \right]_n\!\!\sim$$

Nylon 6

Stabilization of Polyamides

$$\sim CH_2\!-\!C\!-\!N\!-\!CH_2 \sim$$
$$O\cdots Cu\cdots O$$
$$\sim CH_2\!-\!N\!-\!\!\overset{\|}{C}\!-\!CH_2 \sim$$

Polyamide is stabilized with the copper complex shown above.

Antioxidant	Advantages	Drawbacks
Cu salts/ Iodide	Good contribution to long term thermal stability at aging T > 150°C. Very affective at low dosage levels	Leaching in contact with water and solvents. May cause discoloration Dispersability in substrate is critical
Aromatic Amines	Good contribution to long term thermal stability.	Strong discoloring properties Need to be used at high concentrations
Phenolic Antioxidants	Good contribution to long term thermal stability at T < 150°C Good color performance	Inferior long term thermal stability at T > 150°C compared to Cu salts/Iodide system

37

Lithography using iodonium compounds

Photopolymerization catalyst

A photoacid generator is a photosensitizer which generates acid by exposure to light. It is used as a photoresist (photosensitive composition) during the photolithography process in the production of semiconductor devices, printed circuit boards, and LCD (Liquid-Crystal Display) panels. The upper diagram shows the basic lithography process. The photoresist consists of either a negative type, where the pattern of the irradiated area remains, or a positive type where the irradiated area is removed after the development process.

A photoacid generator is comprised of two parts, one that absorbs irradiated light and one that generates acid. A typical photoacid generator is ionic sulfonium salt made up of cations and anions or onium salt such as iodonium salt. An iodonium salt compound is used both as a photosensitizer and photoacid generator for both positive and negative types. An example of a photoacid generator used as a positive-type photosensitizer for chemically amplified resist is shown in the middle diagram. Acid generated by light acts as an acid catalyst, eliminating the protective group and making the resist resin component insoluble. In other words, acid which was generated in the exposed area is catalyzed, causes an elimination reaction of the protective group of the resin component, making the resin component soluble for the alkaline developing agent (tetramethylammonium hydroxide). In the semiconductor photolithography process, this mechanism enables the developing agent to leave the resin component of the insoluble unexposed area and dissolve the resin component of the exposed area, thereby forming a fine pattern.

On the other hand, in the case of negative photoresist, the photoacid generator is used as a cationic photopolymerization initiator. The iodonium compound generates a very strong Lewis acid by light, accelerating the crosslinking hardening reaction of the epoxy novolac resin, making it insoluble. Characteristics of this method are (1) not subjected to hardening inhibition by oxygen, (2) little volumetric shrinkage during hardening due to ring-opening polymerization, and (3) superior properties such as hardened material (cohesiveness and adhesiveness to the base material) [37].

Summary Box

• Iodonium compounds generate acid through photolysis.

• A photoacid generator is used to make a polymer insoluble or soluble.

Principles of Photolithography

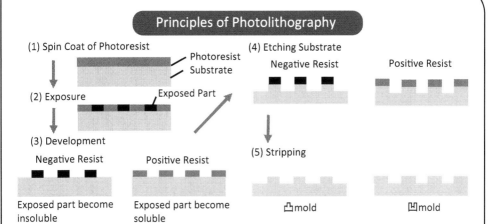

(1) Spin Coat of Photoresist

Photoresist
Substrate

(2) Exposure

Exposed Part

(3) Development

Negative Resist

Exposed part become insoluble

Positive Resist

Exposed part become soluble

(4) Etching Substrate

Negative Resist

Positive Resist

(5) Stripping

凸mold

凹mold

Patterning by Photoresist

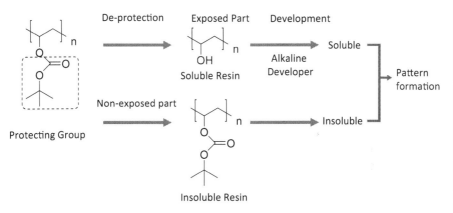

Photoresist : Photo acid Generator + Resin

De-protection

Exposed Part

Development

Soluble Resin

Alkaline Developer

Soluble

Protecting Group

Non-exposed part

Insoluble Resin

Insoluble

Pattern formation

IodoniumCompounds : Photo Acid Generator

$X=CF_3SO_3, C_4F_9SO_3$

By Courtesy of Sanyo Chem. Ind. Ltd.

38

84

Iodine compounds used in fire extinguishing agents

Iodide trifluoromethane

Recently, iodine has begun to be used as a component in fire extinguishing agents. In order for combustible substances to burn, four factors are required: namely, a combustible substance, air (oxygen), sufficiently high temperature, and unrestrained chain reaction. To extinguish a fire, you need only to remove one of these four factors.

There are three methods to extinguish a fire: the cooling method in which water is poured on the combustible substance, lowering its temperature, the smothering method in which the oxygen supply is cut off, and the negative catalyst material method in which the chemical chain reaction is disrupted.

Gas fire extinguishing agents include nitrogen-based fire extinguishers, carbon dioxide-based fire extinguishing agents, and halon-based fire extinguishing agents. Of these, nitrogen and carbon dioxide extinguishing agents are of the smothering type. On the other hand, halon-based extinguishing agents are of the catalyst type. Conventionally, the chemical used for halon-based fire extinguishing agents has been halon 1301 (bromotrifluoromethane), but production has been discontinued since this chemical is high in GWP (global warming potential) and is an ozone-depleting substance.

Subsequently, iodide trifluoromethane (CF_3I) has been developed as a new gas-type fire extinguishing agent [38a,b,c]. The boiling point for CF_3I is $-22.5°C$. It is a colorless nonflammable gas. CF_3I is comprised of three bonds including a very weak bond between carbon atom and iodine atom and a very strong bond between carbon atom and fluorine atoms. As shown in the table, the flame extinction concentration (concentration in the air required to extinguish the fire) differs little from halon, but atmospheric life span is short (0.005 years) and GWP is extremely low (0.4). These properties are promising as a new type of fire extinguishing agent. For example, CF_3I is already used as a fire extinguishing agent for oil tanks in the Netherlands. CF_3I also being considered for use in international airlines and air forces.

Furthermore, CF_3I is also highly expected to be used as a halon substitute for cover gas in magnesium alloys used in mobile terminals and computer parts, as well as in the power sector for insulators and circuit breakers [38d].

Summary Box

- Iodide trifluoromethane is a fire extinguishing agent with low GWP.

Three Elements of Combustion

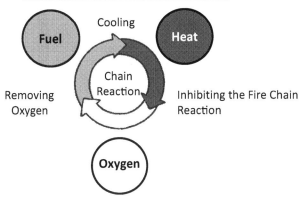

Type of fire extinguishing agent	Inert gas		Halogen based gas	
Fire Fighting Principles	Dilution of oxygen concentration		Inhibition of fire chain reaction	
Gas	N_2	CO_2	CF_3Br (Halon)	CF_3I
Anti-inflammatory Concentration (vol%)	33.6	22.0	2.9	3.0
Global Warming Potential	-	1.0	7140	0.4
Atmospheric Lifetime (Year)	-	-	65	0.005

Production Process of CF_3I

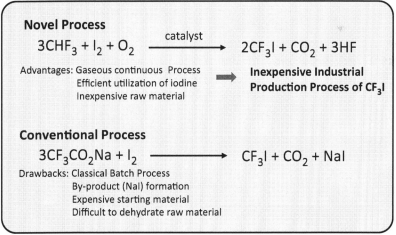

Novel Process

$$3CHF_3 + I_2 + O_2 \xrightarrow{\text{catalyst}} 2CF_3I + CO_2 + 3HF$$

Advantages: Gaseous continuous Process
Efficient utilization of iodine
Inexpensive raw material

➡ **Inexpensive Industrial Production Process of CF_3I**

Conventional Process

$$3CF_3CO_2Na + I_2 \longrightarrow CF_3I + CO_2 + NaI$$

Drawbacks: Classical Batch Process
By-product (NaI) formation
Expensive starting material
Difficult to dehydrate raw material

By Courtesy of New Energy and Industrial Technology Development Organization (NEDO).

39

Iodonium salt for three-dimensional printers

86

Polymerization catalyst

Up to now, most people think a printer only prints on paper. Recently however, 3D (three-dimensional) printers have appeared which can create 3D models (solid objects). By laminating materials with many layers based on 3D data, a solid object can be produced. Lamination methods are roughly divided into the fused deposition modeling (FDM) method, optical modeling method, and power sintering method.

The FDM method involves laminating heat-soluble layers, one by one. This type of 3D printer is now available at low cost and is becoming common.

On the other hand, the optical modeling method is the oldest type of 3D printer. This technology was invented in 1980 by Hideo Kodama (Nagoya Municipal Industrial Research Institute, Japan) [39a], and put into commercial application by 3D Systems Corporation in 1987. This method uses liquid resins (photo-curing resins) which harden when exposed to UV rays. A tank filled with photo-curing resins is irradiated by a UV laser to create the layers. When one layer is complete, the modeling stage sinks one layer, and another layer is laminated. After many layers the model is created. At this point, an iodine compound, which is a photoacid generator, is used as a polymerization catalyst. The photoacid generator is photosensitive, functioning to generate acid with light irradiation, and is an ionic iodonium salt ($Ar_2I^+X^-$) comprised of cations ($+$) which absorb the irradiated light and the anions ($-$) which are the source of the acid. As shown in the reaction formula, irradiated light is absorbed efficiently and strong acid (HX) is generated. The stronger the generated acid, the higher the cation polymerization activity. From the generated acid, monomers such as epoxy and vinyl ester are hardened by polymerization (reaction formula). Photoacid generators must have (1) stable properties, (2) high solubility in solvents and monomers, and (3) high storage stability in a solution blended with polymerizable monomers [39b].

Summary Box

- Photo-curing resins can be hardened with iodonium salt.

- 3D models are produced by cation polymerization.

Principles of 3D Printer (Photolithography)

UV Light

UV Light

3D Printed Object

Liquid Photocurable resin

Fabrication Platform moves down

Completion

Photo-acid Generator for 3D Printer

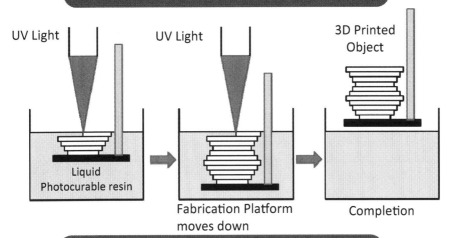

$$Ar_2I^+ X^- + RH \rightarrow ArI + HX + ArR$$

Strong Acid is generated from iodonium compounds by UV Light.

3D Printed Object

Glossary

Cation Polymerization: A type of chain growth polymerization in which a cationic initiator transfers charge to a monomer.

40

Iodine can form iodides with most metals in the periodic table. Metal iodides are produced by the direct reaction between iodine and the metal, the reaction between metal hydroxide and iodine, hydroiodic acid and metal or metal oxide, or by the metal exchange reaction between another metal iodide, etc. Metal iodides prepared this way are used as reactants or accelerants in the synthesis of new substances by chemical reaction. Among the various iodides used in organic synthesis, such as magnesium iodide, indium iodide, samarium iodide, and titanium iodide, the extremely useful samarium iodide (SmI_2) is introduced [40a].

SmI_2 is a green-colored solid generated by thermal decomposition of samarium iodide (III)(SmI_3), and has a melting point of 520°C. SmI_2 is prepared in the laboratory as follows: Thoroughly dry a round bottomed flask and flush with argon. Add samarium metal, iodine crystals, and dried tetrahydrofuran (THF). Stir vigorously at room temperature. When SmI_2 is generated, the color of the solution will turn from orange to blue, finally becoming dark blue. A dark blue THF solution (0.1 M) is commercially available. The characteristic of the reaction using SmI_2 is the ability to monitor the progress of the reaction through color change (loss of the dark blue color).

SmI_2 is a strong reducing agent that reacts rapidly with water to produce hydrogen. In an organic reaction, it can reduce halides, ketones, esters, and sulfoxides. A typical SmI_2 reaction is the Barbier reaction which produces tertiary alcohol from ketone and alkyl halide. There are reports on the intramolecular reaction producing a five-membered ring or a six-membered ring. Furthermore, pinacol coupling stereoselectively proceeds with a high yield [40b].

As mentioned above, metal iodides are important in this modern age.

Metal iodides are the new reactants

88

Samarium iodide

Summary Box

- Reaction of samarium iodide can be monitored.

- Reactions are stereoselective and provide a high yield.

Metal Iodide-Mediated Reactions

High-yielding & Stereoselective Reaction

The reaction can be monitored by the color change.

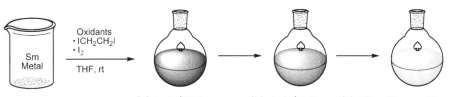

Sm Metal

Oxidants
• ICH$_2$CH$_2$I
• I$_2$

THF, rt

(a) SmI$_2$/THF
Blue

(b) SmI$_2$/H$_2$O
Wine red

(c) After the reaction
Whitish yellow

X   X= Halogen

Dehalogenation

SmI$_2$

Reduction

Dealkoxidation
Desulfonation

Y=OR, S(O)$_n$R

Barbier Reaction

SmI$_2$

THF, r.t.

H$^+$

89~98% (n=2, 3)

Pinacol Coupling

SmI$_2$, HMPA

THF, −78~0°C

54%

41

Hypervalent iodine oxidants are extremely safe

Oxidation reactions are one of the most basic chemical change reactions and are an important industrial process.

Regarding reaction design, the effect on the environment and efficient use of resources must be carefully considered. Trivalent organic hypervalent iodine reagents have similar reactivity to heavy metal oxidants such as highly toxic lead (IV), mercury (II), and thallium (III). However, hypervalent iodine reagents have relatively low toxicity and are easy to handle. Expectations for such environmentally friendly reactants are high.

Many studies up to now have reported on the various reactivities of trivalent hypervalent iodine compounds, as represented by phenyliodine diacetate (PIDA) and phenyliodine bistrifluoroacetate (PIFA). However, trivalent hypervalent iodine compounds are stoichiometric oxidants, producing a monovalent iodobenzene as a byproduct. Recently, as demand for high quality and high purity in the chemical synthesis of drugs, agrochemicals, electronic materials, etc., has grown, removal of the byproduct iodobenzene, which is produced in large quantities, must be addressed in order to achieve commercial application. In response, development of a recyclable hypervalent iodine oxidant is underway. First, a new recyclable hypervalent iodine (III) reagent which introduces four PIDA or PIFA into one molecule with adamantane or methane as the nucleus has been developed. When used in an alcohol oxidation reaction, ketone or aldehyde is produced at high yield. Furthermore, after this reaction is completed, the byproduct iodobenzene may be collected simply by filtering the reaction solution. In addition, by reoxidizing iodobenzene with meta-chloroperoxybenzoic acid, iodobenzene can once again be an active species and be reused in oxidation reactions [41a].

Moreover, creation of a resin-supporting PIDA was also attempted. After the oxidation reaction, poly [4-(diacetoxyiodo) styrene] is collected by filtering, and can be reused and recycled by reoxidization with peracetic acid [41b].

Summary Box

- A highly safe oxidant with low toxicity.

- Hypervalent iodine oxidant can be recycled.

Environmentally friendly reactant

Hypervalent Iodine Oxidizing Agent

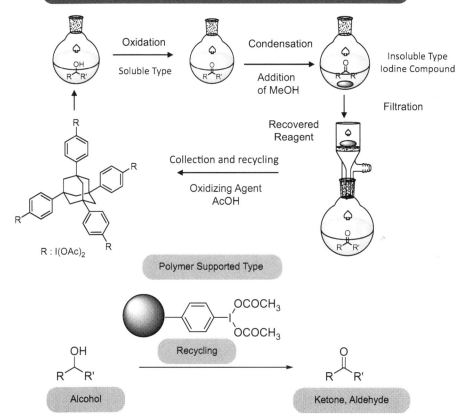

PIDA

PIFA

Koser's Reagent

Dess-Martin Reagent

Hypervalent Iodine Oxidizing Agent

Oxidation / Soluble Type

Condensation / Addition of MeOH

Insoluble Type Iodine Compound

Filtration

Recovered Reagent

Collection and recycling

Oxidizing Agent AcOH

R : I(OAc)₂

Polymer Supported Type

Recycling

OH
R   R'

Alcohol

O
R   R'

Ketone, Aldehyde

Glossary

Hypervalent iodine: Hypervalent iodine has three, five, and seven valences more than one valence.

42

Precise polymerization with iodine compounds

92

Living polymerization

Polymers which have been synthesized by additional polymerization such as polyethylene, PVC (polyvinyl chloride), polyvinyl acetate, polyacrylic ester, etc. are abundantly around us. However, in the process of normal additional polymerization, along with the initiation reaction and growth response, side reactions referred to as transfer and termination reactions occur, resulting in a mixture of polymers of varying molecular weight. However, according to a precise polymerization method called living polymerization, only the initiation reaction and growth response occurs, without any side reactions. Living polymerization shows the following characteristics: (1) molecular weight increases in proportion to the rate of polymerization; (2) molecular weight of the generated polymer can be controlled based on the charged molar ratio or rate of polymerization of the monomer and initiator; (3) polymers of basically the same molecular weight with a narrow molecular weight distribution can be obtained; (4) all generated polymer ends have an initiator section introduced, to allow functional transformation.

In particular, iodine-based compounds are excellent as initiators and catalysts for living radical polymerization and allow the production of structurally controlled polymers with various functions. Specifically, as shown in the lower right diagram, terminal-functional, block, graft, and star polymers and specially structured polymers can be synthesized.

Living radical polymerization can be applied to various monomers such as metacrylate, acrylate, styrene, and acrylonitrile. On the other hand, living polymerization in cation polymerization is said to be difficult. However, when alkylvinyl ester is polymerized using an initiator which combines hydrogen iodide and iodide or zinc iodide, the molecular weight is controlled, resulting in polymers with an extremely narrow molecular weight distribution, proving that living cation polymerization is possible [42].

Furthermore, an environmentally friendly polymerization method combining iodine compounds and organic catalysts has recently been studied, and development of highly functional color materials is now underway.

Summary Box

- Molecular weight can be controlled using living polymerization.

- Functional transformation is possible with an iodine-based polymerization initiator.

Living Radical Polymerization

Addition Polymerization

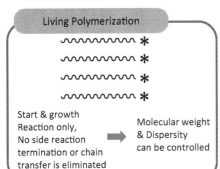

Polymerization Catalyst

monomer

Active Site

Catalyst Site

Growing Polymer

Living Radical Polymerization

Living Polymerization

~~~~~~~ *
~~~~~~~ *
~~~~~~~ *
~~~~~~~ *

Start & growth
Reaction only,
No side reaction
termination or chain
transfer is eliminated

➡ Molecular weight
& Dispersity
can be controlled

Conventional polymerization method

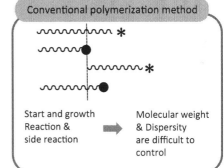

Start and growth
Reaction &
side reaction

➡ Molecular weight
& Dispersity
are difficult to
control

Polymerization by Iodine Compounds

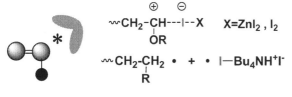

$\overset{\oplus}{\sim}CH_2-CH---I--X$ $X=ZnI_2 , I_2$
       OR

$\sim CH_2-CH_2 \cdot + \cdot I-Bu_4NH^+I^-$
     R

Cation
Polymerization

Radical
Polymerization

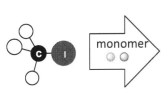

monomer

AB Block Copolymer

ABA Block Copolymer

Graft Copolymer

End functionalized polymers

Star Polymer

43

94

Methods to synthesize organic iodine compounds

Iodinating agents

Iodinated aromatic compounds which bind iodine atoms in the nuclei of aromatic compounds are in great demand as intermediates in many fields such as pharmaceutical products, agrochemicals, and electronic materials. Various iodinating agents are used in the synthesis of aromatic compounds. The following reactants are typical iodinating agents.

Iodine monochloride is an interhalogen produced from chlorine and iodine, and an inorganic compound, represented by the chemical formula ICl. ICl comes in stable α type and unstable β type crystals. The melting point for both types is near normal temperature (27°C) and both are reddish brown. Based on the difference in the electronegativity of iodine and chlorine, iodine monochloride acts as the supply source of iodide ion. These reactants are used in the synthesis of intermediates for the X-ray contrast agents (ATIPA).

One compound formerly used as an organic iodinating agent is N-iodosuccinimide (NIS). 1,3-Diiodo-5,5′-dimethylhydantoin (DIH) is an iodinating agent synthesized from iodine monochloride and dimethylhydantoin.

DIH is an alternative to molecular iodine commonly used in the iodination reaction and provides iodo compounds in high yield and in a regioselective manner compared to NIS. DIH does not have sublimability as molecular iodine, is a low toxic, pale-yellowish solid, and is easy to handle. It is a pale-yellowish to brownish crystalline powder with a melting point of 198°C. Furthermore, dimethylhydantoin, which is generated after reaction, can easily be removed by washing with water. DIH, in the presence of sulfuric acid, reacts quickly under normal temperatures with aromatic compounds, providing the corresponding iodo compounds.

N-iodosaccharin is an electrophilic iodinating reagent, with higher reactivity and selectivity than NIS. For example, in the iodination reaction of aromatic compounds such as phenols and anilines, it does not react with the hydroxyl or amino group, and as a result, introduction of iodine to the aromatic ring at a high yield is possible [43].

Summary Box

- Iodine monochloride is used for the synthesis of contrast media.

- Diiodo-dimethylhydantoin is highly reactive.

Iodination Reagent

N-iodine saccharin

N-iodosuccinimide (NIS)

1,3-diiodo-5,5' - Dimethyl hydantoin (DIH)

ICl

Iodine Monochloride

Iodination by Iodine monochloride

Intermediate of X-ray Contrast Media

ATIPA

Iodination by DIH

Agrochemicals Insecticide

X=1 Precursor
X=2 Flubendiamide

Governments around the world have worked hard to stop counterfeit money. The manufacturing technology of paper currency and that of counterfeiting is a never ending cat-and-mouse game.

In recent years, due to developments in electronic copying technology, counterfeit notes which can hardly be distinguished from authentic ones are in circulation.

In response, various techniques, such as micro-characters, special luminous ink, letterpress printing with deep impressions, watermarks, holograms, etc. are implemented as anticounterfeiting measures.

In addition, low UV wavelength light and magnetic head scanners have been developed to detect counterfeit bills. However, in reality, having many of such devices and systems does not guarantee success. Some of these techniques require expensive machinery, are extremely bulky, or are mechanically complicated.

Subsequently, a simple counterfeit detection pen was developed. When a line is drawn on a counterfeit note printed with a household or a laser printer, the ink turns blackish-brown. Based on the basic principle of detecting the starch content on the note using the iodine starch reaction, the authenticity or fabrication of the note can be detected.

A genuine note is fiber-based (linen or cotton, etc.) and does not contain starch, and does not show any reaction. This pen can be used to detect counterfeit notes, not only for Japanese yen, but also most currencies in the world (US$, Euros, Chinese yuans, British pounds, Korean wons, etc).

Ink used in the counterfeit detection pen is composed of a potassium iodide solution with a mixture of several types of alcohol.

Generally, paper is produced using starch as a sizing agent. Counterfeit notes made from easily obtainable copying paper contain starch, and the bright yellowish-brown ink instantly forms a dark brown or grayish blue-purple complex, changing into a blackish color. With an authentic note, the components evaporate within 24 hours and can be safely used.

Using this pen in a cash transaction with a suspicious customer right in front of one will serve as a strong deterrent.

It is thought that this invention will be influential at airports, major cities where foreign currencies are circulated, in makeshift shop sites, and public institutions.

Counterfeit Detecting Pen

6

Iodine is needed to maintain health

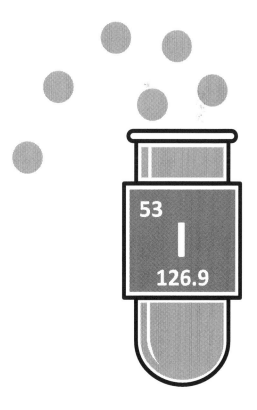

44

The role of iodine in pharmaceuticals

Pharmaceuticals which contain iodine

Halogens (fluorine, chlorine, bromine, iodine) play an important role in the field of pharmaceuticals. The list of halogen-containing pharmaceuticals approved in Japan is shown in the table. Chlorine compounds are most common, followed by fluorine compounds, iodine compounds, and bromine compounds, in that order. Chlorine has been used in the medical field for many years and the introduction of chlorine atoms is a basic structural conversion in pharmaceuticals. On the other hand, with the development of fluorine chemistry, fluorine has an atomic size close to hydrogen, a stable C–F bond, and is now used widely in pharmaceuticals. However, bromine compounds have a comparatively unstable C–Br bond, are highly toxic, and are the least used as pharmaceuticals.

With this background, examples of pharmaceuticals containing iodine atoms, the largest of stable halogens, are as follows: First, typical drugs include the diagnostic drug Iopamidol and Iohexol, X-ray contrast agent. They have 5-amino-2,4,6-triodo-isophthalic acid (ATIPA) as their common mother framework, are chemically stable, and are not subject to biological metabolism. Various hydrophilic substituents are introduced to the side chain and drugs with increased water solubility are used as injections. Amiodarone is an anti-arrhythmic agent used in the treatment of refractory and fatal arrhythmia such as ventricular fibrillation and tachycardia which cannot be treated as other arrhythmia. However, it is accompanied by side effects such as interstitial pneumonia, thyroid function disorder, and cornea deposit formation, and is a difficult-to-use drug.

Idoxuridine is an antivirus ophthalmic solution and is generally used as a therapeutic drug for keratitis occurring resulting from inflammation due to viral infection. Levothyroxine is the oldest iodine-containing drug used as a thyroid hormone agent. There are two types of thyroid hormones, namely triiodothyronine (T3) and thyroxine (T4). T3 activity is said to be several times stronger than T4 (see Section 17) [44].

Summary Box

- Iodine is indispensable in X-ray contrast agents.

- Thyroid hormones are comprised of iodine.

Halogen Containing Drugs

Halogen	Number	Characteristics
F	95	Excellent stability in vivo
Cl	122	Lead compounds of the drug design
Br	13	Toxicity is strong, not used very often
I	20	X-ray contrast agents, Disinfectant

Anti-cancerAgent

antianxiety agents

antianxiety agents

Fluorouracil

Clorazepate
dipotassium

Bromazepam

Iodine Containing Drugs

Levothyroxine (thyroid hormone) 1949
Indication: hypothyroidism, goiter

Idoxuridine (an antiviral agent) 1961
Indication: herpes simplex virus type 2

Amiodarone (antiarrhythmic agent) 1961
Indications: recurrent arrhythmia

Iopamidol (X-ray contrast agent) 1970
Indication: arteriovenous angiography

45

100

Iodine for preoperative sterilization

Povidone iodine ②

Sterilizing agents commonly used in hospitals include benzalkonium chloride, chlorohexidine gluconate, and povidone iodine. Among these, povidone iodine has the longest history of use. Povidone iodine comes in forms such as 10% povidone iodine solution, 7.5% povidone iodine scrub, 10% povidone iodine ethanol solution, 0.5% povidone iodine ethanol hand rub formulation, 10% povidone iodine gel, and 5% povidone iodine cream. Their many uses are shown in the diagram.

Povidone iodine is effective for gram-positive bacteria, gram-negative bacteria, tubercle bacilli, fungi, viruses, and *Clostridium* bacteria, covering a broad antimicrobial spectrum. It is an excellent antiseptic that is hypoallergenic and with relatively few side effects. It can be applied to surgical sites, wounds, as well as mucosa such as the oral cavity and vagina. It is also effective against AIDS (acquired immune deficiency syndrome) and hepatitis B viruses. When povidone iodine is applied to the skin to form a film, a prolonged sterilizing effect can be achieved. However, since it evaporates relatively quickly and loses effect, it is inferior to other iodine agents.

Povidone iodine is a water soluble complex in which iodine is bonded with the carrier PVP. Effective iodine contained in 1 g of povidone iodine is 100 mg. For example, a 10% povidone iodine solution has 1.0% effective iodine.

Povidone iodine maintains a state of equilibrium in the solution and as the free iodine concentration in the solution decreases, iodine is gradually released. Hence, when povidone iodine solution is diluted to around 0.1%, the retention force of the carrier is the lowest and the free iodine concentration is the highest (~25 ppm). However, since povidone iodine is easily absorbed into the body when used on burn sites, the vagina, and oral cavity, side effects such as thyroid metabolic abnormality may occur when used for an extended period time or over a large area. Caution is needed regarding its use [45].

Summary Box

- Iodine sterilizers are effective against bacteria and viruses.

- They are hypoallergenic with few side effects.

Formulation of Povidone Iodine & Application

Povidone Iodine Formulation (Conc.)	USE
Water solution (10%)	Surgical site skin, mucosa of surgical site, mucous membranes wound site, burned skin surface, infection skin surface
Scrub (10%)	Finger-skin, surgical site skin
Ethanol solution (10%)	Surgical site skin
Ethanol scrub (0.5%)	Hand, finger
Gel (10%)	Wound site skin, mucous membranes, burned skin surface
Cream (5%)	Vulva, vaginal
Gargle (7%)	Oral wound, oral cavity

Formulation of Povidone Iodine & Application

Rate of Dilution	Effective Iodine Conc. (ppm) (Titrated with sodium thiosulfate)	Free Iodine (ppm)
Undiluted	10,000	1
1/10	1000	10
1/100	100	25
1/1000	10	8~9
1/10,000	1	1

●———● : Povidone Iodine (PVP-I)

○———○ : PVP-I in 80 Vol % Ethanol

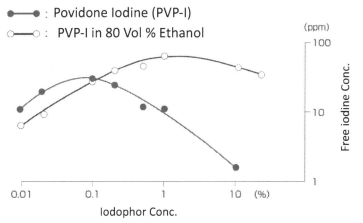

By Courtesy of Yoshida Pharm.

46

102

Iodine compounds as an antiprotozoan remedy

Specific remedy for amoebic dysentery

Amoebic dysentery is an infectious disease of the digestive tract found around the world, but particularly in Asia, Africa, and South America. Every year, approximately 100,000 people worldwide lose their lives to this disease.

In Japan, this infectious disease is often seen among overseas travelers. However, cases of mass infection in welfare facilities and among male homosexuals have been reported, and cases of the disease show an upward trend. Amoebic dysentery spreads through drinking water or ice, eating fruit, vegetables or raw meat that has been infected with *Entamoeba* in the fecal matter of an infected person; 10%–20% of those infected manifest symptoms. Normally, symptoms such as diarrhea, mucous and bloody stool, tenesmus, lower abdominal pain, and discomfort during defecation appear 2–4 weeks (range of few days to few years) of ingesting the pathogen. A typical example is mucous and bloody stool resembling strawberry jelly, and cycles of improvement and deterioration at intervals of a few days to few weeks.

The international standard remedy for amoebic dysentery is a non-iodine compound, metronidazole. However, iodine compounds are widely used overseas as a specific remedy. These compounds are thought to manifest a sterilizing effect by extracting metal from intracellular metalloenzyme using a metal chelating action. An example of this is to include 5,7-diiodo-8-hydroxyquinoline, which can be easily produced through iodination of iodine monochloride and use it as an inexpensive amoebic dysentery remedy. Dosage is 650 mg, 3 times/day, 20 days.

Another compound is 5-chloro-8-hydroxy-7-iodoquinoline. This drug was widely sold as an intestinal drug in Japan under the name of Quinoform. However, it was found to be a causative factor for SMON (subacute myelooptico-neuropathy) disease and the production of this "drug with a shady history" was discontinued in 1970. However, it is still sold in some countries as a specific remedy for dysentery [46a,b].

Summary Box

- Diiodo-hydroxyquinoline and quinoform are effective against amoebic dysentery.

- Overdosing and long-term taking of quinoform led to the occurrence of SMON.

Amoebic Dysentery Remedy

Iodine Compounds

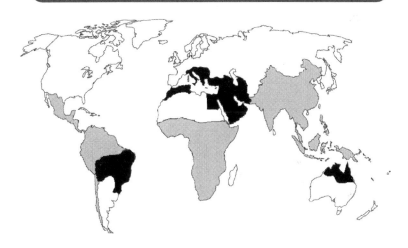

5,7-diiodo 8-hydroxy quinoline

Quinoform

Non-iodine Containing Compounds

Metronidazole
(Frazier)

Amoebic Dysentery

Infection rate of residents (>1%) pathogenic Entamoeba histolytica

Infection rate of residents (>1%) non-pathogenic Entamoeba histolytica

Glossary

SMON is hazard caused by Quinoform, an antiseptic, prescribed for the treatment of diarrhea and other bowel symptoms. Its overdosing and long-term taking led to the occurrence of SMON, which is an iatrogenic disease of the nervous system leading to a disabling paralysis, blindness and even death.

47

Iodine-based contrast agents for brain and heart examinations

104

Iodine-based contrast agents shield X-rays

X-ray contrast agents are widely used for cardiovascular and cerebrovascular examinations and other detailed examinations performed in hospitals. In most cases, iodine-based drugs are used. As iodine shields X-ray transmission, it provides an excellent contrast to other tissues.

X-ray contrast agents are largely divided into ionic (hyperosmolar) contrast agents such as diaatrizoate, and nonionic contrast agents (hypotonic) such as ipamidol and iotrolan. The basic structure of an X-ray contrast agent is triiodobenzene. The first contrast agents to be introduced in the 1950s were ionic contrast agents. Ionic contrast agents were associated with side effects such as vascular endothelial damage derived from the hyperosmolar properties of ionic compounds, blood–brain barrier dysfunction, thrombosis, and thrombophlebitis.

In the 1980s, a water soluble and nonionic side chain with a hydroxyl or ether structure was introduced to the 1-, 3-, and 5-position on the benzene ring by means of an amide bond. Characteristics of nonionic contrast agents are: (1) no charge, (2) the hydrophobic group (iodine) is covered with a hydrophilic side chain, weak protein binding capacity and enzyme inhibiting action, and minimal effect on biomembrane function, and (3) low osmotic pressure. From this, safety has improved, and side effects such as nausea/vomiting, hives, mucosa swelling, increased breathing resistance, and effects on the cardiovascular system are significantly lower than ionic contrast agents.

The iodine content ratio of non-iodine contrast agents (see the diagram) that are most widely used in urography, angiography, and contrast-enhanced CT is 300–370 mg/mL, and the osmotic pressure of the formulation is low, approximately 3 times that of blood (6 times for ionic). Non-iodine contrast agents are currently used in normal examinations at many facilities. When injected into the veins, a heat sensation is felt, and the body feels comfortably warm. Although rare, minor side effects include hives, sneezing, nausea, etc. [47]

Summary Box

- Nonionic X-ray contrast agents have a high level of safety.

- X-ray contrast agents are hydrophilic and dissolve well in water.

Inspection by X-ray contrast agent

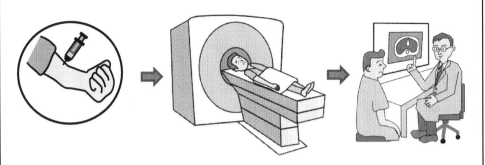

Diatrizoic Acid	Iopamidol	Iotrolan

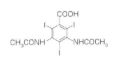 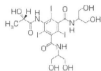 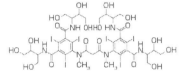

 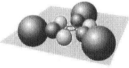 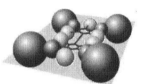

Ionic Agent

 Iodine
 Side Chain

Nonionic Agent
(Monomer Type)

Nonionic Agent
(Dimer Type)

By Courtesy of Bayer.

(See color insert.)

105

	Iodine Conc. (mgI/mL)	Osmolality (mOsm/kg H_2O)	Viscosity (mPa/s,37℃)
Blood	-	290	-
Sodium Diatrizoate	292	1050	-
Iopamidol	300	644	4.5
Iotrolan	300	291	8.1

48

106

Stable iodine protects against radioactive iodine

Iodine tablets distributed around the nuclear power plant

With the nuclear power plant accident due to the Great East Japan Earthquake as a catalyst, investigation and measures which give priority to the development of legislation and guidelines regarding a radiation emergency medical system have been implemented.

If an accident occurs at a nuclear power facility, various radioactive substances may be released. One of such substance, radioactive iodine, accumulates in the "thyroid" since the human body cannot differentiate between radioactive and stable iodine. As a result, when an individual is exposed to large quantities of radioactive iodine, he/she is at greater risk for thyroid abnormalities such as thyroid cancer. However, if a stable iodine tablet (potassium iodide) is taken beforehand, accumulation of radioactive iodine in the thyroid gland can be avoided, and the carcinogenic risk may be reduced despite exposure to radioactive iodine [48a]. Incidentally, the most effective time to consume an iodine tablet is immediately before radioactive iodine is ingested into the body. If the tablet is taken at this time, accumulation of radioactive iodine may be inhibited by 90% or more (see the table). However, caution is needed since the effect of iodine tablets decreases if too much time elapses before exposure to radioactive iodine. When radioactive iodine is absorbed or ingested, it rapidly accumulates in the thyroid and is metabolized as an organic iodine compound. This organic compound remains in the thyroid gland for a long period of time, triggering local radiation damage, such as thyroid nodules and cancer.

Prior to the accident at the nuclear power plant, stable iodine was prescribed for the treatment of goiter associated with hyperthyroidism, sputum expectoration associated with chronic bronchitis and asthma, and tertiary syphilis. However, due to additions in efficacy and effect, oral administration of potassium iodide at the dosage, shown in the upper right-hand table, is now possible, in order to prevent or alleviate internal exposure of radioactive iodine to the thyroid [48b,c]. However, as of 2015, the only potassium iodide being sold domestically is the potassium iodine 50 mg pill from Nichi-Iko Pharmaceutical Company, Japan.

Summary Box

- The ingredient for stable iodine tablet is potassium iodide.

- The mechanism to suppress accumulation of radioactive iodide in the thyroid.

Stable Iodine Administration and Effectiveness

Timing of KI Administration (100 mg of KI)	Effectiveness against ^{131}I
24 hours before exposure	70%
12 hours before exposure	90%
0 hours before exposure	97%
3 hours after exposure	50%
6 hours after exposure	non

Age Group	Iodine
Birth – 1 month	12.5 mg
Over 1 month to 3 years	25 mg
Over 3 - 13 years	38 mg
Over 13	76 mg

Radioactive Iodine Uptake Inhibition by Stable Iodine

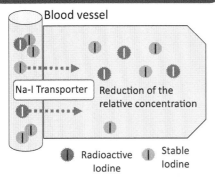

Blood vessel

Na-I Transporter

Reduction of the relative concentration

Radioactive Iodine Stable Iodine

Stable Iodine Pills

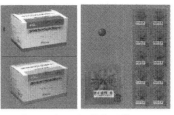

Potassium iodide Pills
By Courtesy of Nichi-Iko Pharm.
(See color insert.)

Internal Exposure to Radioactive Iodine

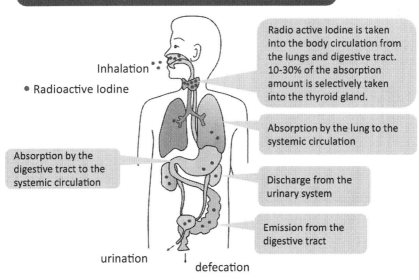

Inhalation

• Radioactive Iodine

Radio active Iodine is taken into the body circulation from the lungs and digestive tract. 10-30% of the absorption amount is selectively taken into the thyroid gland.

Absorption by the lung to the systemic circulation

Absorption by the digestive tract to the systemic circulation

Discharge from the urinary system

Emission from the digestive tract

urination   ↓ defecation

49

108

Soon after the Great East Japan Earthquake and the subsequent Fukushima No. 1 Nuclear Power Plant disaster on March 11, 2011, iodine became well publicized. However, it was not until sometime later that a distinction between stable iodine and radioactive iodine was made.

The molecular weight of radioactive iodine is 131, whereas that of stable iodine is 127. Stable iodine is an essential element for living organisms, regulating growth and metabolism as a thyroid hormone in humans and animals. On the other hand, radioactive iodine has been used for treatment of cancer and as a diagnostic drug.

Typical radioactive iodine isotopes include iodine 123, iodine 125, and iodine 131 (see the table). There are 15 radioactive iodine drugs, constituting one-third of all radioactive drugs. Iodine 123 has a half-life (13.2 hours) and γ ray (159 keV) energy suitable for diagnostic imaging. Iodine 123 is used for 12 diagnostic radiopharmaceuticals including ioflupane [^{123}I]. Iodine 125 has a long half-life of 59.4 days and emits weak γ ray energy (27.5 keV), and is suitable for radiation treatment. For example, an iodine 125 seed (^{125}I encapsulated in a 5 mm long, 1 mm diameter titanium capsule) is sold commercially. It is embedded into the focus of a prostate cancer patient using a dedicated needle.

Iodine 131 may be used both as diagnostic and therapeutic drugs. γ rays (364 keV) are used for diagnosis and β rays for treatment. Sodium iodide [^{131}I] is a radioactive drug which has been used in treatments since ancient times. When administered orally, 30% of the radiation is accumulated in the thyroid gland. As a result, it is used in the treatment of hyperthyroidism and goiter. Tositumomab [^{131}I], a monoclonal antibody with iodine 131 added, has raised the 10-year survival rate of malignant lymphoma by 50% [49a,b]. Hence, not all radioactive iodine should be considered bad.

Radioactive iodine as therapeutic and diagnostic drugs

Radioactive iodine is not always the bad guy

Summary Box

- One-third of radiopharmaceuticals are of iodine compounds.

- Radioactive iodine compounds can diagnose illnesses.

Pharmaceuticals Containing Radioactive Iodine

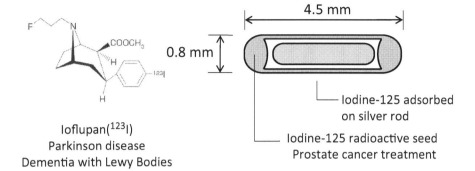

Ioflupan(^{123}I)
Parkinson disease
Dementia with Lewy Bodies

4.5 mm

0.8 mm

Iodine-125 adsorbed
on silver rod

Iodine-125 radioactive seed
Prostate cancer treatment

IMP(^{123}I)

Diagnosis of Regional
Cerebral Blood Flow

Iomazenil(^{123}I)
Diagnosis of epilepsy

Half Life & Pharmaceutical Applications of Radioactive Iodine

Radioactive Iodine	Half Life	Applications
^{123}I	13.3 Hr	Diagnostic Agent
^{125}I	59.4 Day	Therapeutic Agent
^{131}I	8.03 Day	Oral Medicine

Ioflupan(^{123}I) Injection
(See color insert.)

By Courtesy of Nihon Medi-Physics Co., Ltd.

50

Iodine lithium batteries used in the medical field

Cardiac pacemakers

A pacemaker is a medical device to monitor and treat bradyarrhythmia. Through a lead connected to the main device, electrical signals from the heart are monitored 24 hours a day. When the heartbeat of the wearer needs to be regulated, it sends out an electrical stimulation.

The main outer component of the pacemaker is made of titanium, which is a sturdy metal. Inside, batteries supply power for an extended period of time to a control circuit, which serves as the unit's brain. In fact, the batteries for a pacemaker are unique, and contain iodine.

Iodine lithium batteries were developed by Catalyst Research in the United States in 1974. Iodine batteries have lithium as the anode, a mixture of iodine and poly-2-vinyl pyridine (P2VP) as the cathode, and a lithium iodide as the solid electrolyte.

As pacemakers are embedded into the body as shown in the diagram, a high level of safety and long life are required. The high level of safety of an iodine lithium battery lies not only in the use of lithium iodide as the inorganic solid electrolyte, but in the combination of cathode, anode, and solid electrolyte. In other words, in the cell reaction of an iodine lithium battery, ionization reaction occurs at the anode, and lithium iodide formed in the cathode becomes the separator of the cathode and anode.

Therefore, in the unlikely event that the solid electrolyte which doubles as a separator is damaged and the anode and cathode short circuits, a chemical reaction in the anode and cathode instantly forms lithium iodide (electrolytes), ensuring the safety of the lithium iodine battery [50a,b].

The battery life of iodine lithium batteries is 10 years, but they are usually changed every 5–7 years. According to the national survey in Japan, over 57,000 new and replaced pacemakers were implanted in 2010 alone.

Summary Box

- Iodine lithium batteries are safe.

- Iodine lithium batteries have a long battery life.

Mechanism of Pacemaker

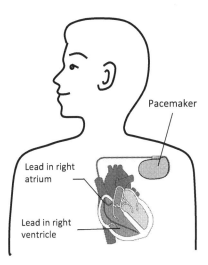

Pacemaker

Lead in right atrium

Lead in right ventricle

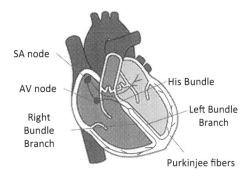

SA node

AV node

His Bundle

Right Bundle Branch

Left Bundle Branch

Purkinjee fibers

Electrical signals start at the SA node, causing atria contraction, and then move on to AV node.

A pacemaker is a medical device which uses electrical impulses, delivered by electrodes contracting the heart muscles, to regulate the beating of the heart.

Structure of Pacemaker

Pacemaker Body

Electrode

Battery

Circuitry

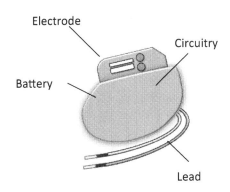

Lead

Lithium Iodine Battery

Electrode

Electrode

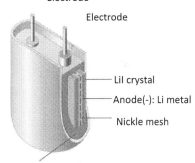

LiI crystal

Anode(-): Li metal

Nickle mesh

Cathode(+): I_2 complex

Cathode	$I_2(s) + 2e^- \longrightarrow 2I\text{-}(LiI)$
Anode	$2Li(s) \longrightarrow 2Li^+(LiI) + 2e^-$
Cell reaction	$2Li(s) + I_2(s) \longrightarrow 2LiI(s)$

51

112

An iodine inclusion complex for bedsore treatment

Cadexomer iodine

With a rapidly aging population and the increase in elderly persons who are bedridden, the number of decubitus (bedsores) cases has increased and is a major concern for those involved in the healthcare field. Decubitus is caused by poor blood circulation of subcutaneous tissue, resulting in a circulatory disorder as oxygen and nutrients do not reach the extremities, and necrosis of the skin and tissue occurs. Persons who are bedridden are often physically weak and unable to personally change their lying position for long periods of time and are extremely likely to develop "bedsores." The buttocks and areas of bone protrusion, as shown in the diagram, are most vulnerable.

Here, a wound-healing agent Iodosorb (cadexomer iodine formulation), which is used in the treatment of decubitus and skin ulcers (burns), is introduced. Iodosorb (powder form) became available in Japan in 1993, and Iodosorb ointment in 2001. Cadexomer iodine solution is a formulation encapsulating iodine in a cadexomer substrate, synthesized by crosslinking the dextrin obtained from hydrolysis of potato starch with epichlorohydrin, and then carboxylated. Cadexomer, similar to the polyacrylic acid used in disposable diapers, has a 3D structure. When exposed to water, the polar carboxyl groups repel one another due to Coulomb's force, forming a molecular chain that spreads out like a fishing net structure. Water is retained in each mesh opening. As a result, iodine retained in the cadexomer is gradually released to produce sustained sterilization.

In addition, cadexomer works to absorb, adsorb, and clean any exudate, viscous necrotic tissue, phlogenic material, bacteria, etc. A cadexomer iodine formulation has greater sterilizing performance than other similar agents. For example, it is strongly effective against MRSA.

The sterilizing effect of iodine is influenced by the surrounding pH. With the cadexomer iodine formulation, acidity is maintained and effects are particularly long lasting [51a,b].

Summary Box

- Cadexomer iodine is used for bedsores.

- Iodine-based sterilizers are multidrug resistant.

Location of Pressure Sores

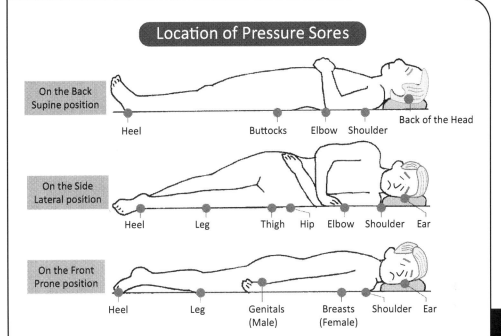

On the Back
Supine position

Heel — Buttocks — Elbow — Shoulder — Back of the Head

On the Side
Lateral position

Heel — Leg — Thigh — Hip — Elbow — Shoulder — Ear

On the Front
Prone position

Heel — Leg — Genitals (Male) — Breasts (Female) — Shoulder — Ear

Dosage Form of Cadexomer Iodine

(See color insert.)

Chemical Structure of Cadexomer Iodine

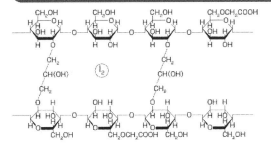

By Courtesy of Smith & Nephew Japan

Alexander the Great became King of Macedonia in the fourth century BC at the age of 20, and successively conquered Greece, Asia Minor, Egypt, and Persia during his 12-year rule. He built a large empire extending from Egypt to the west of India, and is famous as a military leader. There are many stories and theories regarding how and why Alexander the Great died at the young age of 32 in 323 BC, such as malaria or bacterial infection, infection from drinking river water, wounds from the battlefield, and even assassination. The answer to that mystery has not yet to be unveiled. But one theory proposes that Alexander the Great was poisoned by calicheamicin, one of the few natural products containing iodine.

In 1981, calicheamicin was isolated from bacteria existing in caliches (limestone) in Kerrville, Texas. The cytotoxicity of calicheamicin is 3000 times stronger than general anticancer drugs, and was first marketed in 2000 as Gemtuzumab ozogamicin, a molecular target drug for nonsolid carcinoma acute myeloid leukemia. The carbohydrate chain in calicheamicin recognizes and bonds with the DNA (deoxyribonucleic acid) sequence, and the core of the enediyne generates biradicals which cause DNA cleavage. Currently, the iodine atoms are thought to determine positioning of the attack site.

An American researcher cited the similarities between the symptoms of Alexander the Great immediately before his death and the symptoms of calicheamicin poisoning. In addition, Mt. Chelmos in Peloponessus, Greece is calcareous, and the waters of the Mavroneri River may have been contaminated with calicheamicin.

Iodine-containing natural products may rewrite ancient history.

Calicheamicin

7

Iodine for vegetable production and livestock breeding

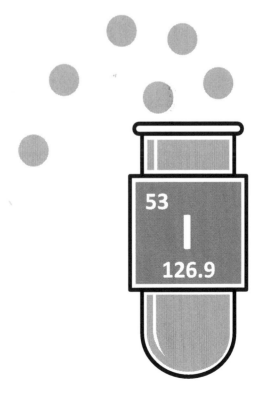

52

116

Iodine compounds for livestock feed additives

Essential trace minerals for cows and sheep include zinc, copper, manganese, iron, iodine, cobalt, selenium, etc. However, the amount of these minerals in livestock feed changes depending on the soil and natural environments of the country.

Japanese soil contains a sufficient amount of iodine. However, in Western countries, iodine compounds such as sodium iodide, potassium iodate, and ethylenediamine dihydroiodide (EDDI) are used as livestock feed additives.

Foot rot is a communicable bacterial disease which affects the hooves of sheep. If just one sheep is infected, the whole flock will instantaneously be infected. To prevent the spread of foot rot, daily observation and prevention are essential, and when foot rot is discovered, immediate treatment is imperative. To prevent foot rot and its spread, regular hoof trimming (once every 2 months), disinfection by foot bath (a bath of copper sulfate or zinc sulfate solution once a week), a clean breeding environment, etc. are essential. In addition, antibiotic treatment and disinfection using sterilizers are also carried out.

However, foot rot can be prevented by providing trace minerals, in particular iodine and zinc. Based on test results, both minerals have been found to be highly effective.

Furthermore, evidence suggests a correlation between iodine deficiency in cows with goiter, actinobacillosis, actinomycosis, cervical actinomycosis, and foot rot. Based on test results on beef cattle, a preventative effect for foot rot has been indicated. Administration of EDDI at a dose of 10–15 mg/animal day^{-1} is effective in the prevention of foot rot. According to the U.S. Food and Drug Administration (FDA), the regulated maximum dose of EDDI for dairy cattle is 10 mg/animal/day, 500 mg/day. This is to set to limit the transfer of iodine to milk to 500 ppb or less [52].

As seen from the above, not only humans but also livestock require iodine to maintain their health.

Ethylenediamine dihydroiodide

Summary Box

- Livestock feed additives.

- Preventive drug for goiter and foot rot.

Iodine Based Feed Additive

Trace Mineral Salt with Iodine

Trace Mineral Salt with Iodine

Chemical Name	Chemical Structure
Potassium Iodide	KI
Potassium Iodate	KIO_3
Calcium Iodate	$Ca(IO_3)_2$
Etylenediamine Dihydroiodide	$NH_2CH_2CH_2NH_2 \cdot 2HI$

Iodized Salt for Cattle

Prevention of Foot Rot in Cattle

Hoof wall gets thick and overgrown.

Hoof wall cracks

Inflammation in Cattle feet (Laminitis)

The erosion of tissue between the sole of the toe and the hard outer hoof characterizes foot rot.

Foot Rot Preventing Method

Ouch!

- Proper hoof trimming
- Disinfect the trimming instruments
- Regular foot bathing (Disinfection)
- Provide good drainage to all areas in pastures and paddocks .
- Stop grazing in the pasture where foot rot occurred.
- Early detection of infection animals
- Isolate affected animals
- **Supplemental iodine (see the table above)**

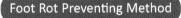

53

Iodine-based sterilizers used during milking

118

Prevention of udder inflammation

Mastitis in dairy cows results in great economic loss for dairy farmers. Mastitis is categorized into two types depending on the causative bacteria. To effectively prevent mastitis, countermeasures according to the cause must be implemented.

Infectious mastitis is often transmitted through milk drawn from udders infected with the bacteria during milking, and through contaminated milking equipment or the hands and fingers of the milker. In addition, pathogens may remain on the udder after milking so disinfecting the udder and killing the bacteria by dipping is required. This disinfection method is called postdipping.

On the other hand, environmental mastitis occurs when the causative bacteria exist on the floor or straw mulching of the cow shed, in cow excrement or urine, or on the cow's body. These bacteria infect the udder between milkings and normally remain on the udder until milking time. As a result, predipping, namely disinfecting the udder immediately before milking, is effective.

Currently, the most widespread dipping agent in Japan is iodine-based (iodine, potassium iodide, nonoxynol iodine, povidone iodine, etc.), followed by a fatty acid-based dipping agent (glycerin fatty acid ester). Other agents used by farmers include chlorhexidine gluconate and sodium hypochlorite. The most commonly used iodine-based product is called iodophor, and various hydrophilic additives are included to stabilize iodine.

Dairy farmers are also instructed to wash the udder thoroughly before milking in order to remove any iodine residue. This is to avoid any bad health effect to the human consumer (iodine overdose, etc.) [53a,b]. Following the footsteps of dairy farmers overseas, pre-dipping is becoming widespread in Japan.

Summary Box

- Iodine-based sterilizers used before and after milking.

- Care is given to avoid residual iodine in milk.

Antimicrobial & Bactericidal Activity of Typical Commercial Disinfectants

Type of Agent	Active Ingredient	Fungi			Virus	
		Viable Bacteria	Fungus	Spore Bacteria	Enveloped	non Enveloped
Biguanide	Chlorhexidine gluconate	○	△	×	△	×
Phenol-based	Ortho-dichlorobenzol	○	○	×	△	×
Inverted soap	Benzalkonium chloride	○	△	×	△	×
Chlorine-based	Sodium hypochlorite	○	○	△	○	○
Iodine-based	**Nonoxynol-iodine**	○	○	×	○	○
	Iodine glycine complex	○	○	×	○	○

○ High antimicrobial △ Moderate to Weak antimicrobial × Not antimicrobial

Procedure of milking

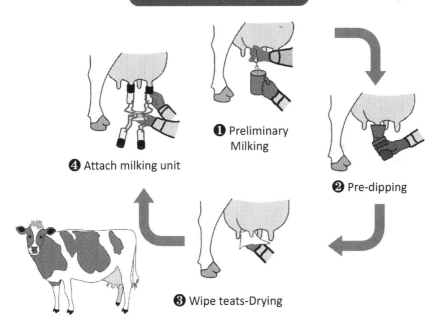

❹ Attach milking unit

❶ Preliminary Milking

❷ Pre-dipping

❸ Wipe teats-Drying

Glossary

Mastitis is an inflammation in the breast tissue (also sometimes called "milk fever"), often caused by a blocked milk duct that hasn't cleared.

54

Iodine is useful as a weed killer

Iodine-based agricultural chemicals and herbicides

As the world population increases, securing adequate food supply is an important issue. In addition, from a labor-saving viewpoint of farm work, there is a demand for safe and effective agricultural chemicals. Herbicides have the image of being harmful. However, in recent years, effective and safer agricultural chemicals which use halogens have been developed. Compared to chlorine and fluorine, agricultural chemicals containing iodine are few. However, there are several chemical agents which take advantage of the properties of iodine atoms.

Ioxynil is a widely used herbicide developed in 1963 by May & Baker to spray annual broad-leaved weeds such as chickweed, henbit, and shepherd's purse when growing monocotyledons such as wheat variety, corn, and onions.

Ioxynil is absorbed by the stems and leaves of the weeds, inhibiting photosynthesis and respiration, causing the plant to wither and die. Bromoxynil, which contains bromine instead of iodine, has similar action. However, ioxynil is effective over a wider variety of weeds and has a longer effect [54a,b,c].

Iodosulfon-methyl is a chemical agent developed in 1999 by AgrEvo USA. It is primarily used as a herbicide for wheat and corn. Compared to conventional chemical agents, a lower dose is effective on more than 50 types of broad-leaved weeds. Regarding its mechanism, acetolactate synthase is inhibited, preventing the synthesis of branched chain amino acids such as leucine, isoleucine, and valine. Biosynthesis of normal proteins is blocked, growth is discontinued, and the plant withers and dies. Compared to similar chemical agents not containing iodine, not only is effectiveness high but it decomposes rapidly, reducing environmental load [54d,e].

Halogens can accept electrons from electron-releasing elements such as nitrogen, oxygen, and sulfur (see Section 64 for halogen bond). Among the halogens, iodine has the strongest halogen bonding interaction, and new agricultural chemicals using iodine are currently being designed.

Summary Box

- Long-used, iodine-containing herbicides.

- Synthesis of branched amino acids is inhibited.

Iodine Based Herbicide

N≡━⟨benzene ring with I substituents⟩━OCO(CH₂)₆CH₃

Ioxynil (IOX)

IOX is a contact-type and nitrile-based herbicide. IOX inhibits photosynthesis by binding to the D1 protein, reduces the production of ATP and NADPH necessary for the survival of weeds, and leads the weed to death.

Iodosulfuron-methyl sodium (IMS)

IMS is sulfonylurea herbicides to control broad-leaved weeds incereals. IMS is based on inhibition of the enzyme acetolactate synthase (ALS) which is responsible for biosynthesis of free branched chain amino acids: valine, leucine, and isoleucine.

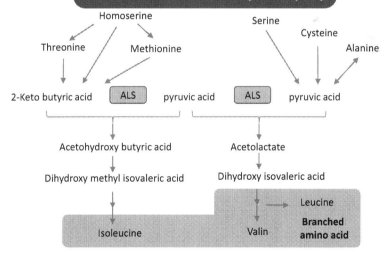

Inhibition of Acetolactate Synthase(ALS)

Homoserine

Threonine — Methionine

Serine — Cysteine — Alanine

2-Keto butyric acid — ALS — pyruvic acid — ALS — pyruvic acid

Acetohydroxy butyric acid — Acetolactate

Dihydroxy methyl isovaleric acid — Dihydroxy isovaleric acid

Leucine

Branched amino acid

Isoleucine — Valin

55

122

Unlike antibacterial agents in the medical field, agricultural sterilizers are chemicals which destroy or inhibit the growth of pathogenic microbes on flowers, vegetables, and fruit. Without agricultural sterilizers, the amount of vegetables and fruit harvested would significantly decrease. Examples of agricultural sterilizers and disinfectants which contain iodine are Proquinazid [55a], Iodocarb (lipid and cell membrane synthesis inhibitor) [55b], and Benodanil (succinate dehydrogenase inhibitor) [55c]. Of these, Proquinazid is a relatively new agricultural sterilizer developed by DuPont in 2005.

Powdery mildew is a disease where white specks first appear on the leaves and gradually spread to the entire plant until it becomes white as if dusted with wheat flour. If left unattended, the surface of the leaves becomes white, photosynthesis is blocked, inhibiting growth. A low dose of Proquinazid is effective against powdery mildew. It is also effective on unaffected leaves, with long lasting effect. The mechanism is still unclear, but inhibition of the formation of appressoria, without inhibiting spore germination and the growth of the germ tube has been reported.

Another agent for powdery mildew is called Quinoxyfen, but its mechanism differs from that of Proquinazid. Quinoxyfen inhibits the function of GTP-binding proteins in intracellular transduction. Currently, the substitution effect of pyridopyrimidine in the development of Proquinazid is presently being examined in detail. Reactivity with 7-iodine, a bromine substitute was high, but was significantly low with the three-dimensionally and electronically similar methyl group, trifluoromethyl group, phenyl group, nitro group, and cyano group. Quinozolinone derivative showed a similar tendency, and when iodine was introduced to the 6- or 6- and 8-positions, the highest level of activity was attained.

In this way, we can enjoy delicious fruit and vegetables thanks to iodine-based sterilizers.

Summary Box

- Iodine sterilizers are effective against powdery mildew at low doses.

- Iodine-based agricultural chemicals prevent diseases in vegetables and fruits.

Iodine Based Disinfectant

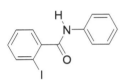

Proquinazid

Benodanil
Succinate dehydrogenase
inhibitor

I——≡≡——CH$_2$OCONHBu-*n*

Troysan (IPBC)
Lipid and cell membrane
synthetic inhibition

Powdery Mildew

Infected plants display white powdery
spots on the leaves and stems.

COOH   COOH
H–C–H   H–C   H
H–C–H   H–C   H
COOH   COOH

Succinate dehydrogenase
inhibition

The agricultural productivity associated
with iodine-based chemicals.

Vegetable safety comes from iodine

Iodine-based agricultural chemicals and pesticides

Halogens (fluorine, chlorine, bromine, iodine) have played an important role in the development of agricultural chemicals. In particular, chlorine has been widely used as an component of pesticides such as DDT (dichlorodiphenyltrichloroethane) and BHC (benzene hexachloride) (see the diagram). However, the use of chlorine-based agricultural chemicals have been discontinued in developed countries due to safety concerns. In recent years, the introduction of fluorine has shown to significantly change drug efficiency and properties, and the use of fluorine-based agricultural chemicals containing fluorine atoms and the trifluoromethyl group (CF$_3$) have rapidly increased due to their high level of safety. Overall, 30% of all agricultural chemicals presently used are said to be fluorine-based. On the other hand, because iodine is comparatively costly compared to other halogens and in limited supply, hence there are fewer cases of use in agricultural chemicals.

Flubendiamide is a genuine, domestically produced pesticide developed by Nihon Nohyaku Co., Ltd. and registered in 2007. One characteristic of Flubendiamide is that it has 1 iodine atom and 7 fluorine atoms in a molecule and is highly effective against green caterpillars that eat cabbage and lettuce [56a].

Flubendiamide binds to the ryanodine receptor in muscle cells in insects and disrupts the calcium ion concentration. As a result, muscle fibers constrict abnormally and the insect immediate stops eating and starves to death. Mammals also have ryanodine receptors, but homology greatly differs between mammals and insects, and the portion that binds to Flubendiamide is missing in mammals. As a consequence, toxicity to humans and animals is considered low. Unlike existing pesticides, Flubendiamide has a completely different mechanism and is effective against organic phosphorus, carbamate, and pyrethroid insecticide-resistant pests [56b,c]. One characteristic of Flubendiamide is that the interval between spraying and harvest is short. In particular, it can be used on strawberries, cabbage, lettuce, and tomatoes up to the day before harvest. This is highly advantageous compared to many other competitive pesticides.

Summary Box

- Low toxicity to humans and animals and is considered safe.
- Can be used up to the day before harvest.

Insecticides in the Past

DDT

BHC

The Structure and Properties of Flubendiamide

Molecular Formula	$C_{23}H_{22}F_7IN_2O_4S$
Molecular Weight	682.39
Water Solubility	29.9×10^{-3} mg/L (20°C)
Distribution Coefficient	$\log_{10}Pow = 4.20$ (24.9±0.1°C)

The Mode of Action of Flubendiamide

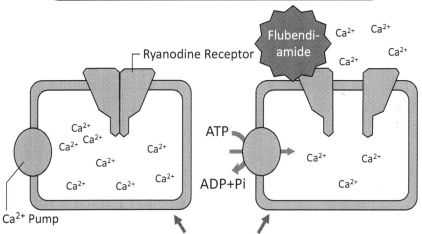

Ryanodine Receptor

Flubendi-amide

Ca^{2+}

ATP

ADP+Pi

Ca^{2+} Pump

Endoplasmic reticulum in insect muscle cell

Flubendiamide disrupts proper muscle function in insects and therefore represents a novel, unique mode of action. These characteristic symptoms are induced by Flubendiamide through the activation of ryanodine-sensitive intracellular calcium release channels (ryanodine receptors, RyR).

Iodine sterilization system for agriculture

126

A sterilization system known as the ISAN system has been developed in Australia. A patent for this system has already been obtained in Australia and the United States. First, an iodine solution is prepared using iodine contained in a tank. Solubility of the iodine in water is dependent on water temperature, but is approximately 250–300 ppm.

With the ISAN system, iodine concentration is detected through electrodes to automatically control the amount of iodine solution to be added, and the ppm level of the iodine concentration of the spray can be freely set.

As mentioned in previous sections, iodine is effective against most bacteria and viruses and is naturally effective as an agricultural sterilizer. Examples of practical application are as follows: First, sterilization of a hydroponics culture for tomatoes. Iodine is also used to sterilize the irrigation water used to raise seedlings. In addition, it is used in postharvest sterilizers and disinfectants for fruit and vegetables. The ISAN system is becoming popular in poultry and livestock raising. In Southeast Asia, poultry raising is a thriving business. In this area, the system is used to sterilize drinking water for poultry in a breeding farm, preventing infections and supplementing iodine intake. In Australia and New Zealand, the system has also been approved for the sterilization of eggs. The ISAN system is used widely in disinfecting livestock sheds in dairy farms, sterilization of udders before and after milking, and sterilization of the milking line.

Iodine concentration in Australian and New Zealand soil is low, and iodine deficiency is on a national scale. Therefore, since 2009, iodine has been added to bread. This may be a contributing factor to the relatively relaxed nature of iodine regulations in the field of agriculture and livestock breeding.

Apart from the field of agriculture, the ISAN system is used to sterilize swimming pools, although this is not approved in Japan. Uses for this system are expanding to include sterilization of hospital facilities and waste disposal landfill sites [57].

Iodine concentration control system

Summary Box

- Iodine solution concentration can be automatically regulated.

- Iodine sterilization system is becoming popular in agriculture fields.

ISAN System Applications

Nurseries

Hydroponics

Dairy

Poultry

ISAN System

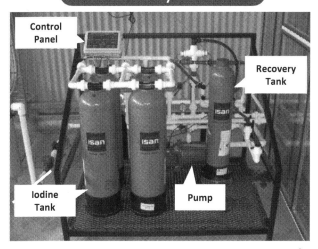

Control Panel

Recovery Tank

Iodine Tank

Pump

By Courtesy of IOTEQ.
(See color insert.)

58

128

Rainfall made possible with iodine

Artificial rainfall

According to the United Nations, by 2025, two-third of the world's population will face water shortages. The situation of inland countries is especially critical. Even in Japan, areas with dense population are faced with the possibility of a water shortage. In this situation, studies on artificial rain are being conducted in various parts of the world. Artificial rain has a long history, and American meteorologist Charles Hatfield is said to have successfully produced artificial rain over 100 years ago.

Currently, silver iodide is most widely used in artificial rain, and this method was discovered by American meteorologist Bernard Vonnegut [58a]. As shown in the diagram, this method, called "seeding," involves dispersing silver iodide in rain clouds in order to create minute ice grains that stimulate the production of snow. If the temperature near the ground is above 0°C, the snow melts and falls as rain. This method was established around 1946 when studies on artificial rain began. Thereafter, silver iodide was dispersed efficiently in clouds using airplanes and rocket artillery. In China and Russia, artificial rain is used as a countermeasure for water shortage. Especially artificial rain is very much useful in agriculture at the time of drought or water scarcity.

In Japan, there are two artificial rainfall devices, one located in Okutama Town, Tokyo, and the other in Kofu City, Yamanashi Prefecture. These devices burn silver iodide and acetone mixture to artificially produce the nuclei of ice crystals, and inject the smoke into rain clouds, causing rainfall. By operating this device, an approximate 5% increase in precipitation can be expected (see the right hand bottom lower diagram) [58b].

Conversely, this method was used to produce "fine weather" at the opening ceremony for the Beijing Olympics in 2008. The weather forecast for that evening in Beijing was "thunderstorm." However, a total of more than 1000 rockets were fired into the rain clouds in Beijing city and its surroundings several hours before the ceremony, Silver iodide was dispersed to produce rainfall in advance, a strategy called "artificial rain suppression." As a result, by the time of the opening ceremony, Beijing city miraculously had fine weather.

Summary Box

- Silver iodide is dispersed in rain clouds to cause rainfall.

- Artificial rain is useful in agriculture at the time of drought.

Principles of Artificial Snow-rain Technology

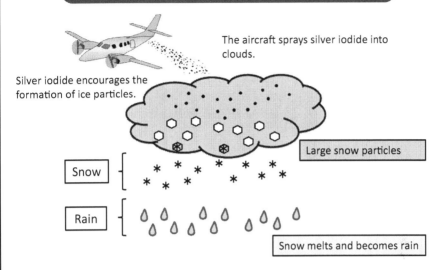

The aircraft sprays silver iodide into clouds.

Silver iodide encourages the formation of ice particles.

Large snow particles

Snow

Rain

Snow melts and becomes rain

Artificial Rain Generator

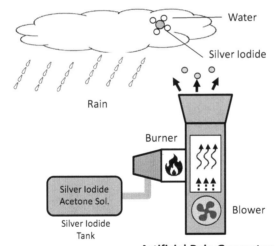

Water

Silver Iodide

Rain

Burner

Silver Iodide Acetone Sol.

Silver Iodide Tank

Blower

Artificial Rain Generator

Glossary

Silver iodide (AgI) is a pale yellow, odorless powder used in medicine especially as an antiseptic, and in photography. The crystalline structure of AgI is similar to that of ice, allowing it to induce freezing in cloud seeding for the purpose of rainmaking. Approximately 50,000 kg/year are used for this purpose[58c].

Urban mines refer to the recycling concept advocated by Professor Michio Nanjyo of the Institute of Mineral Dressing and Metallurgy at Tohoku University, where industrial products accumulated on land are regarded as resources, called "urban mines" and proactive extraction of these resources is attempted. Various parts of a circuit board in an electronic device contain rare metals and rare earths. Recycling of such precious metals using iodine compounds is discussed below.

There are many suggestions regarding recycling methods for rare precious metal resources. For example, a conventional method is to dissolve the precious metals using inorganic acids such as aqua regia, etc. Other proposed measures include the dissolution method that uses alkali cyanide, and a method to dissolve the platinum group by blowing chlorine into hydrochloride acid. However, the dissolution method using inorganic acids such as aqua regia require a strongly acidic solution, which adds a heavy burden on the maintenance and durability of plant facilities. In addition, toxic gases such as nitric oxide are generated during dissolution, posing a danger to the workers, not to mention the extremely burdensome task of absorbing the generated gas and treating the exhaust gas. Since cyanide is a highly toxic substance, extreme caution is required to protect the workers from danger when alkali cyanide is used in the method, and detoxification treatment to dispose of used cyanide is a heavy burden. Furthermore, with the method of blowing chlorine into hydrochloric acid, providing corrosion resistance of the facilities is also an issue.

On the other hand, a number of methods using iodine compounds in rare metal recycling have been reported. For example, a mixture of iodine monochloride and iodic acid has higher solubility for gold and palladium than aqua regia, and can efficiently recycle rare metals. Compared to the abovementioned chemicals, the mixture is safe and disposal of liquid waste is also facilitated. In addition, this solution can also be applied to the recycling of ruthenium and rhodium which have low solubility. Moreover, a method for recycling gold using potassium triiodide has also been reported. We anticipate iodine to further contribute to the development of urban mines in the future.

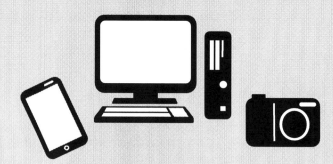

Electrical and Electronic Equipments as Urban Mines

8 Next-generation technology starts with iodine

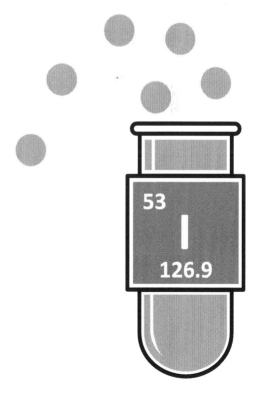

59

132

Iodine is the deciding factor for
next-generation solar cells

Dye-sensitized solar cells

As concerns grow over global warming caused by burning fossil fuels and the safety of nuclear power generation, expectations for photovoltaic cells which use the inexhaustible energy of the sun have increased. In particular, dye-sensitized solar cells (DSSC) have gained attention as low-cost, next-generation solar cells due to their simple structure, inexpensive materials, and manufacturing process. In 1991, Professor Gratzel of the Swiss Federal Institute of Technology in Lausanne, Switzerland, developed DSSC comprised of electrodes where a ruthenium metal complex is adsorbed to porous titanium oxide, iodine-based electrolytic solution, and platinum counter electrodes. This triggered active research worldwide, and now, improvements have been led to a conversion efficiency exceeding 11% [59a,b].

The action concept of DSSC is shown in the diagram. First, the dye absorbs sunlight and becomes excited. Next, the electrons are transferred to titanium dioxide, generating electromotive force. The dye receives the electrons from iodide ions in the electrolytic solution and returns to the ground state. The triiodide ions generated in this process are reduced at the counter electrodes. Through this series of chemical reactions, an electric current is created.

The typical manufacturing process for DSSC is described below. First, a paste containing dispersed, fine titanium oxide particles is applied to a substrate with a transparent conductive film, and then baked (400–500°C) to create a porous titanium oxide film. This film is then immersed in a dye solution so that the dye is adsorbed to the titanium oxide surface. Next, an iodine-based electrolytic solution is injected between the counter electrodes to seal the surroundings. Since the production process of DSSC does not require a vacuum or high-temperature processing, they can be manufactured at a lower cost compared to conventional silicon-based solar cells [59c]. In addition, there is greater freedom in choosing the color and shape design. Moreover, DSSC can be generated in dim morning and late afternoon light as well as by artificial room lighting.

Summary Box

- Low-cost solar cells are easy to manufacture.

- Colorful and flexible solar cells.

Principle of Dye Sensitized Solar Cell (DSSC)

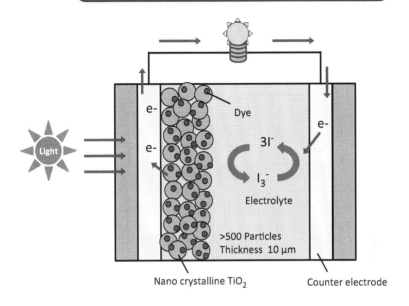

Dye

e-

e-

Light

$3I^-$

I_3^-

Electrolyte

e-

>500 Particles
Thickness 10 μm

Nano crystalline TiO_2 Counter electrode

DSSC Dyes

COON(Bu-n)$_4$

COOH

NCS,,,

Ru

NCS

COOH

COON(Bu-n)$_4$

N-719

COON(Bu-n)$_4$

COOH

NCS,,,

Ru

NCS

SCN

COON(Bu-n)$_4$

Black Dye

Characteristics of DSSC

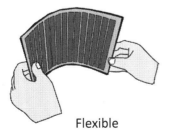

Flexible

Colorful

Image provided by Sony Co.
(See color insert.)

60

Iodine and nuclear power-generated waste heat used to produce hydrogen

134

IS hydrogen production process

In the development of technology to produce hydrogen as a future energy source, the use of waste heat from nuclear power reactors has gained attention.

For direct decomposition of water by heat, the water must reach a temperature of approximately 4000°C. However, decomposition at approximately 800°C is possible using chemical reaction. There are more than 100 thermochemical processes, but in one process, namely the IS (Iodine-Sulfur) process which uses iodine (I_2) and sulfuric acid (H_2SO_4), high hydrogen production efficiency is anticipated. Development of this method to use exhaust heat from high-temperature nuclear gas reactors is now underway [60a].

Basic steps in the IS method are shown in the diagram. The IS method is comprised of three processes, namely sulfuric acid decomposition in which oxygen is obtained through thermal decomposition of sulfuric acid at approximately 850°C, hydrogen iodide decomposition where hydrogen is obtained through thermal decomposition of hydrogen iodide at 400–500°C, and the Bunsen reaction where the iodine, sulfurous acid gas, and water obtained from the above thermal decomposition is converted to hydrogen iodide and sulfuric acid at around 100°C. From this, hydrogen can be produced with just water and thermal energy.

As strong acids such as sulfuric acid and hydrogen iodide are used in the IS method, corrosion-resistant materials are an issue. In particular, in the sulfuric acid decomposition process where sulfuric acid is evaporated at approximately 400°C and decomposed at approximately 650°C, a heat exchanger to conduct heat from the nuclear reactor is needed, and materials which can be meet the parameters are limited to precious metals, glass, and ceramics. Currently, the most promising material is silicon carbide ceramics (SiC), and development of this material for heat exchangers is currently underway [60b,c].

Large-scale hydrogen production is possible by the combination of a high-temperature gas reactor and the IS process, which is expected to meet high future demands [60d].

Summary Box

- Hydrogen is produced with iodine and sulfur.

- Hydrogen is produced using waste heat from nuclear power generation.

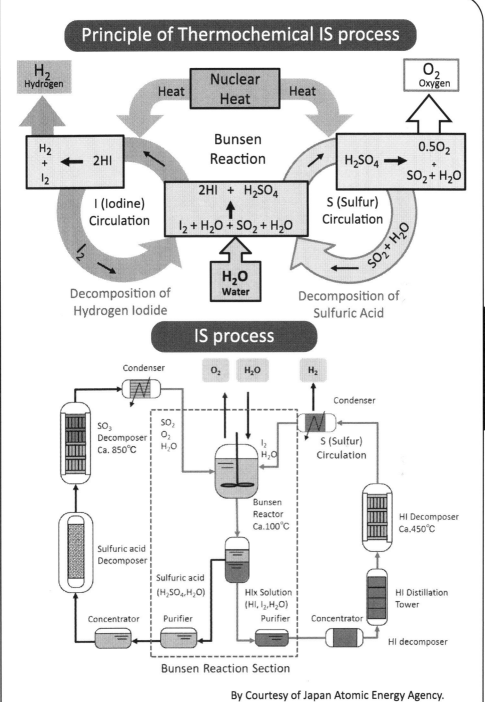

Principle of Thermochemical IS process

H₂ Hydrogen

O₂ Oxygen

Nuclear Heat

Heat ← Nuclear Heat → Heat

Bunsen Reaction

$H_2 + I_2$ ← 2HI

H_2SO_4 → $0.5O_2 + SO_2 + H_2O$

I (Iodine) Circulation

$2HI + H_2SO_4$

$I_2 + H_2O + SO_2 + H_2O$

S (Sulfur) Circulation

$SO_2 + H_2O$

I_2

H₂O Water

Decomposition of Hydrogen Iodide

Decomposition of Sulfuric Acid

IS process

Condenser

O₂   H₂O   H₂

Condenser

SO₃ Decomposer Ca. 850°C

SO₂ O₂ H₂O

I₂ H₂O

S (Sulfur) Circulation

Sulfuric acid Decomposer

Bunsen Reactor Ca.100°C

HI Decomposer Ca.450°C

Sulfuric acid (H_2SO_4, H_2O)

HIx Solution (HI, I_2,H_2O)

HI Distillation Tower

Concentrator   Purifier   Purifier   Concentrator   HI decomposer

Bunsen Reaction Section

By Courtesy of Japan Atomic Energy Agency.

61

136

Iodine for water purification in space craft

Sterilization device for recycled water

The International Space Station (ISS), orbiting at an altitude of 400 km from the earth, is operated under the cooperative efforts of space agencies from the U.S., Russia, Japan, Canada, and Europe, and conducts observations of the earth and space, as well as various studies and experiments related to the space environment. Water replenishment on the ISS is greatly reduced by efficiently reusing wastewater. To reduce the amount of water supply required for manned space exploration in a harsh space environment, the reuse of water is an extremely vital issue. Current replenishment capacity at ISS can only support up to six people, the standard number of crew members.

A special water recycling system is used at ISS to produce pure distilled water from condensed water, the urine of crew members, and their cleaning/washing water. This distilled water, along with other wastewater collected from the spacecraft, is combined to ultimately be processed in the water treatment device to create drinking water for crew members. Iodine adsorption resin is used in this water treatment device. Similar to adding chlorine to household drinking water, iodine treatment is carried out to inhibit the growth of microbes. Iodine is used instead of chlorine because solid iodine adsorption resin is far easier to transport in orbit than chlorine gas.

The iodine adsorption resin used here is a combination of triiodide ions and an anion exchange resin (refer to Section 12). The mechanism of bacteria elimination is that the iodine adsorption resin (positive ion state) draws and adsorbs mold and bacteria in the air and water (negative ion state), and the iodine seeping out from the resin surface sterilizes it instantly. The U.S. Environmental Protection Agency (EPA) authorized the use of this iodine adsorption resin for the drinking water purification system, and this system has been used in all space shuttles up to now. In addition, the Pentagon has also approved its use for water bottles carried by military personnel. In many countries around the world, this iodine adsorption resin is incorporated in water purification systems to procure clean and safe water in cases of disasters and in areas where humanitarian aid is provided [61a,b].

Summary Box

- Easy-to-transport iodine adsorption resin.

- Iodine for water purification at the battlefront and in case of disaster.

Space Station & Water Recovery System

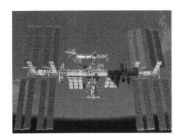

International Space Station

By Courtesy of JAXA/NASA.

Water Recovery System

By Courtesy of NASA/Dimitri Gerondidakis.

ISS Environmental Control and Life Support System (ECLSS)

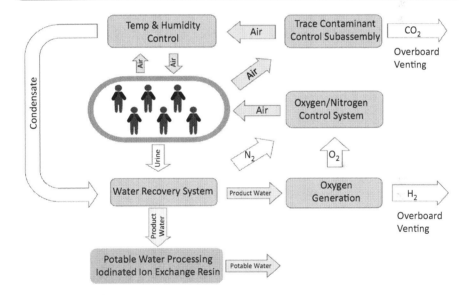

62

Cross-coupling made possible by iodine

138

Hypervalent iodine compound

Cross-coupling refers to a reaction which binds together two carbons with different properties within one organic compound. Negishi and Suzuki cross-coupling are representative reactions. However, costly rare metals are required as a catalyst for these reactions, or extra steps in the synthesis process are required to introduce halogen, boron, and metal into the aromatic ring [62a].

Professor Yasuyuki Kita et al. of Ritsumeikan University developed a metal-free cross-coupling reaction using hypervalent iodine as a catalyst as shown in the diagram below [62b].

Iodine can easily expand in valence to form a trivalent or pentavalent hypervalent iodine compound, exceeding the Octet rule. For example, a trivalent iodine compound has the following structure. The central iodine atom creates a plane with two unshared electron pairs and one σ bond, and a ligand with high electronegativity (L) forms a bond at the apical position orthogonally intersecting that plane. The hypervalent bond in this apical position has a longer bond distance than the σ bond, and has high reactivity because it can easily be cleaved.

By using a hypervalent iodine reactive agent as a catalyst, iodine which is an abundant resource in Japan, the use of costly metal catalysts and processes to activate reaction can be eliminated. In addition, the problem of by-product homodimer formation can be resolved. Furthermore, hypervalent iodine compounds can be used not only in the formerly difficult cross-coupling of heteroaromatic rings and carbon aromatic rings, but also in the selective cross-coupling of heteroaromatic rings with similar properties. Heteroaromatic polymers have superior electroconductivity, transparency, and antioxidative effect. Thus, application as an industrial material has great potential. In the future, cross-coupling using hypervalent iodine will no doubt prove useful in the development of various functional materials such as conductive polymers, liquid crystal materials, and solar cell materials [62c,d].

Summary Box

- Coupling reaction is possible with hypervalent iodine.

- Functional polymers may also be synthesized.

Cross couplings using organometallic compounds

Preactivation

Rare-metal catalyst

R^1

M

M=B, Sn

Pd、Ni、Zn、Cu etc.

R^1 R^2

R^2

R^2

X

X=Br, I

Cross Couplings Using Hypervalent Iodine Reagent

Environmentally Benign Synthesis Methods without Metal Catalyst

R^1 + R^2

Lower cost iodine reagent

Unfunctionalized
aromatic compounds

R^1 R^2

L — Apical Position

Ar — I

L

R—S

+

N
Ph

Hypervalent iodine(III)
reagents

R—S — N
Ph

The I-L bond in the apical position is weaker and longer than
the covalent bond, and highly reactive.

Hypervalent Iodine Reagents

By Courtesy of Tokyo Chem. Ind. Co. Ltd.

63

140

Iodine laser

Laser light generated by chemical reaction

Laser technology involves a method to bring molecules and atoms into an excited state, and then sandwiching them in a resonator (mirror) to extract a light beam.

Normally, electricity or optical energy is used to activate the molecules and atoms. However, there is a special type of laser which uses chemical reaction. Since chemical lasers can achieve higher output than other types of lasers, their applied research has been conducted in nuclear fusion, rocket propulsion, missile defense, etc., since the 1970s.

One example of such research is the chemical oxygen iodine laser (COIL) system. Under COIL, lasers are created in the following sequence. First, potassium hydroxide (KOH) is reacted with hydrogen peroxide (H_2O_2) and chlorine gas (Cl_2) to generate oxygen (O_2) in a metastable state, then iodine gas (I_2) is added in the reactor to bring iodine atoms to an excited state, injecting them from a supersonic nozzle. Iodine gas, through adiabatic expansion, instantaneously turns cryogenic, and radiates a strong laser beam [63a, b]. COIL was developed by the United States for military purposes such as missile interception, and practical application tests have proven successful.

On the other hand, studies on the industrial use of iodine lasers, such as cutting and welding, have been conducted at universities and companies in Japan. Characteristics of an iodine laser are (1) A shorter wavelength of 1.3 µm compared to a carbon dioxide laser wavelength of 10.6 µm, for a larger welding penetration depth. (2) Miniaturization of the optical device in the processed tip can be more easily achieved compared to carbon gas lasers, since an optical fiber guide can be implemented. (3) Energy can be transmitted and distributed over a long distance, making remote control possible, etc. Kawasaki Heavy Industries has successfully carried out high-speed welding at 1 m/min with an iodine laser. In addition, Professor Masamori Endo of Tokai University has recently succeeded in developing a gas phase chemical iodine laser using gases such as trichloramine (NCl_3), hydrogen atoms (H), and hydrogen iodide (HI), with potential use in the removal of space debris [63c].

Summary Box

- Chemical lasers developed for missile interception and the removal of space debris.

- High-speed welding was successfully carried out by iodine laser.

Comparison of Chemical & Solid-state Laser

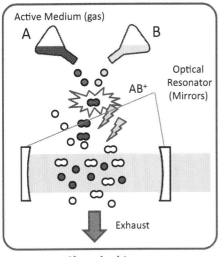

Chemical Laser

Active Medium (gas)

A   B

AB$^+$

Optical Resonator (Mirrors)

Exhaust

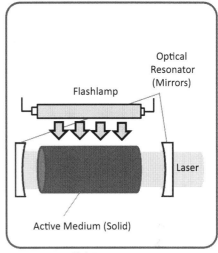

Solid State Laser

Optical Resonator (Mirrors)

Flashlamp

Laser

Active Medium (Solid)

Iodine Laser

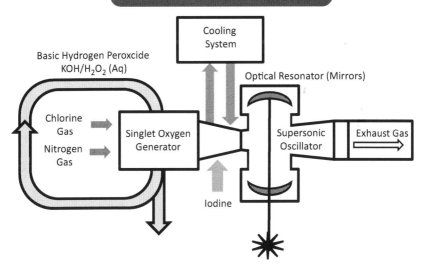

Cooling System

Basic Hydrogen Peroxcide KOH/H$_2$O$_2$ (Aq)

Optical Resonator (Mirrors)

Chlorine Gas

Nitrogen Gas

Singlet Oxygen Generator

Supersonic Oscillator

Exhaust Gas

Iodine

Iodine Laser

By Courtesy of Tokai Univ.

64

Molecular design with halogen bonding of iodine

142

A new concept called halogen bonding (see the diagram) has gained much attention in iodine chemistry, and the First International Conference on Halogen Bonding was held in Porto Cesareo, Italy, in 2014 [64a, b]. All four types of halogens, namely iodine (I), bromine (Br), chlorine (Cl), and fluorine (F), have been theoretically and experimentally proven to be capable of halogen bonding interactions, and that the order of bonding strength is I > Br > Cl > F. Thus, iodine was shown to normally form the strongest interaction.

Development of electronic materials and pharmaceutical designs using these interactions is currently underway. For example, Professor Resnati et al. of Politecnico di Milano is designing a liquid crystal compound by forming a molecular assembly through the halogen bonding interactions of diiodoperfluoroalkane or perfluoro-iodobenzene and pyridine derivatives which have abundant electrons (left center diagram).

In addition, Dr. Imakubo et al. of Nagaoka University of Technology introduced I–N-type halogen bonds to tetrathiafulvalene (TTF)-based organic conductor crystals, and successfully developed a supermolecular organic conductor and a hexagonal crystal-based supramolecular organic conductor (lower left diagram) [64c].

Halogen bonding has also gained attention in the pharmaceutical field. Professor Ho et al. of Oregon State University indicated the importance of interaction between the carbonyl group and hydroxyl group of the receptor protein in the expression of thyroid hormone activity and the O–I-type halogen bond of iodine atoms in thyroid hormones [64d].

In addition, Dr. Roughley et al. of Vernalis verified the structure–activity relationship in the biological activity and strength of halogen bonds in the molecular design of the anticancer drug PD325901 (MEK1 inhibitor), and confirmed the high biological activity of iodine compounds. In this way, consideration of halogen bonds in electronic material and pharmaceutical designs can be said to be the new trend in this field [64e].

Strongest halogen bonding is achieved with iodine

Summary Box

- Halogen bonds arrange molecules.

- Thyroid hormones function through halogen bonds.

Comparison of Hydrogen Bonding & Halogen Bonding

Hydrogen Bonding

$$D—H \cdots A—$$

(D=O、N、C、S) (A=O、N、S)

Halogen Bonding (XB)

$$D—X \cdots A—$$

(X=Cl、Br、I)

(D=O、N、C、S) (A=O、N、S)

: I>Br>Cl>F

XB & Liquid Crystal

$C_8H_{17}O$ —〈benzene〉— CH=CH —〈pyridine〉— N----I —〈C_6F_5〉 (F, F, F, F, F)

XB & Superconductors

XB & Drug Design

Val 127 (MEK1)

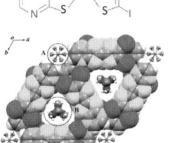

(See color insert.)

Glossary

Halogen Bonding: Noncovalent interaction between a halogen atom (Lewis acid) and a Lewis base.

65

Development of new drugs from marine iodine compounds?

Halogen-containing natural marine products

Many useful biologically active substances have been discovered and isolated from marine organisms. In particular, seaweed, soft corals, and sponges which feed on seaweed are a treasure trove of halogen-containing compounds, with more than 2,400 diverse natural products not found among land organisms identified. These iodine-containing natural marine products manifest interesting biological activities such as antifungal, antibacterial, antivirus, and anti-inflammatory activities. The halogen content ratio in seawater is chlorine 19,350, bromine 67, iodine 0.06, and fluorine 1.3 mg/kg. What is the ratio of natural halogen-containing marine products? Bromine-containing compounds are found most abundantly, followed by chlorine-containing compounds. Iodine-containing compounds are relatively few, and fluoride-containing compounds are seldom found. This is thought to be related to the halogen oxidation reduction potential and cell membrane permeability and is not proportional to the halogen concentration in seawater.

There are only a few (~180) natural marine products that contain iodine, but here are some examples. Lukianol-B is isolated from sea squirts living near the Palmyra Atoll and has aldose reductase inhibitory activity, a causal factor of diabetes. Plako hyperforin is an iodine-containing indole compound, the first natural product to be identified. Iodovulone was isolated from soft coral tsutsuumizuta that live in the coral reef near Ishigaki Island, Okinawa Prefecture, Japan. The chemical structure for tsutsuumizuta differs significantly from mammal-derived prostanoid, and is a new type of prostanoid with strong anti-ulcerative properties. Cytostatic activity against the human leukemia cell line of iodovulone has an IC_{50} of 0.03 mg/mL and is somewhat weaker than chlorovurone. 5-Iodotubercidin has been isolated from red algae in Australia and has a strong adenosine kinase inhibitory effect. It is commonly used as a biochemical reagent.

Iodo-pyriformin is found in natural sponges in the Caribbean and has been verified as having antimalarial activity [65].

Summary Box

- Bioactivity of sea-iodine compounds is gaining attention.

- There are fewer sea-iodine compounds than chlorine compounds.

Marine Iodine Compounds

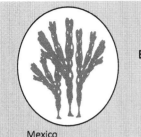

Mexico
Brown algae

Iodo-Filiformin
Antimalarial activity

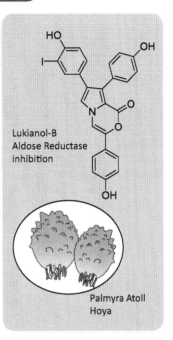

Lukianol-B
Aldose Reductase
inhibition

Palmyra Atoll
Hoya

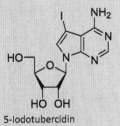

5-Iodotubercidin
Adenosine Kinase Inhibitors

Australia
Red algae

Caribbean Islands
Sponge

Plakohypaphorine
Antihistaminic activity

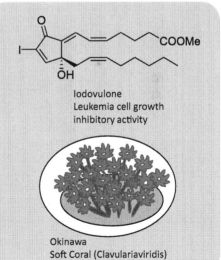

Iodovulone
Leukemia cell growth
inhibitory activity

Okinawa
Soft Coral (Clavulariaviridis)

New Japanese-made solar cells are manufactured with a crystal structure called perovskite and have gained the attention of researchers worldwide. In just 5 years since their development, the energy conversion rate has improved by fivefold and is fast approaching mainstream silicon-based solar cells.

Organic–inorganic hybrid perovskite compounds (MeNH$_3$PbI$_3$) are semiconductor materials used as a light-absorbing layer, and play an important role in perovskite solar cells. High-purity lead iodide (II) is used as an ingredient for perovskite compounds, to produce superior solar cell devices.

Perovskite solar cells do not require a high-temperature or high-vacuum process and can be produced simply by application. Professor Tsutomu Miyasaka of Toin University of Yokohama reevaluated this perovskite crystal as a component. In 2009, he discovered that the thin film on the perovskite crystal for the generator part could function as a solar cell. In 2012, in a joint research project with Oxford University, Professor Miyasaka et al. achieved a 10.9% conversion efficiency and reported their research results in a U.S. academic journal "Science". While the present conversion efficiency is still at the laboratory level, an approximately conversion efficiency of 20%, similar to that of compound-based solar cells which use gallium, has been achieved.

Characteristics of perovskite solar cells are not only that they are easy to produce, but can be produced at low cost. High-temperature or high-vacuum processes are not needed. The solar cells can be produced by merely applying and drying the solution onto a porous titanium oxide on a substrate. While issues regarding stability and practical application must still be resolved, the crystal raw materials are very inexpensive and very promising for next-generation solar cells.

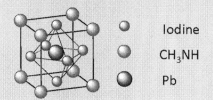

○ Iodine
○ CH$_3$NH
○ Pb

Perovskeit
(See color insert.)

New Japanese-made Solar Cells

References

1. http://www.gasukai.co.jp/iodine/index2.html (accessed December 12, 2016).
2. Lyday P.A., Kaiho T. 2015. *Iodine and Iodine Compounds in Ullmann's Encyclopedia of Industrial Chemistry*, Wiley-VCH, Weinheim.
3. (a) Courtois B. 1813. Découte d'une substance nouvelle dans le Vareck. *Ann. Chim.* 88:304–310.
   (b) Marshall J.L., Marshall V.R. 2009. Rediscovery of elements: Courtois and iodine. *Hexagon Winter* 100(4):72–75. https://chemistry.unt.edu/sites/default/files/users/owj0001/iodine.pdf (accessed December 12, 2016).
4. Kaiho T. 2014. Part II production of iodine: Production process in the past. In: Kaiho T. (ed.), *Iodine Chemistry and Applications*. John Wiley & Sons, Hoboken, NJ, pp. 207–211.
5. Polyak D.E. 2015. *USGS Year Book Iodine 2013*, February. http://minerals.usgs.gov/minerals/pubs/commodity/iodine/myb1-2013-iodin.pdf (accessed December 12, 2016).
6. (a) The Southern Kanto gas field. http://www.gasukai.co.jp/english/gas/index4.html (accessed December 12, 2016).
   (b) The salt concentration in brine. http://www.gasukai.co.jp/english/iodine/index3.html (accessed December 12, 2016).
7. Kaiho T. 2008. Industrial production and Application of Iodine. *IDD Newsletter*, February, pp. 12–14.
8. (a) Lauterbach A. 1999. Production technology of iodine in Chile. *The 2nd Symposium on Iodine Utilization*, October 16, Chiba University, Chiba, Japan, pp. 29–38.
   (b) Lauterbach A., Uber G. 2011. Iodine and iodine compounds. In: *Kirk-Othmer Encyclopedia of Chemical Technology*. John Wiley & Sons, Inc., New York, pp. 1–28.
9. (a) Kaiho T. 2014. Part II. Production of iodine: Recycling of iodine. In: Kaiho T. (ed.), *Iodine Chemistry and Applications*. John Wiley & Sons, Hoboken, NJ, pp. 243–247.
   (b) http://www.godoshigen.co.jp/english/service/iodine/recycle.html
10. Küpper F.C., Feiters M.C., Olofsson B., Kaiho T., Yanagida S., Zimmermann M.B., Carpenter L.J. et al. 2011. Commemorating two centuries of iodine research: An interdisciplinary overview of current research. *Angew. Chem. Int. Ed.* 50:11598–11620.
11. (a) BASF Safety Data Sheet: PVP-Iodine 30/06 Revision date: 2015/08/26 Version: 3.0 BASF Technical Information Povidone Iodine USP August 2010 issue dated June 2008.
    (b) Gottardi W. 1980. Redoxpotential and germicidal action of aqueous halogen solutions. *Zentralbl. Bakteriol. Hyg I Abt Orig B* 170:422–430.
    (c) Gottardi W. 2001. Iodine and iodine compounds. In: Block S.S. (ed.), *Disinfection, Sterilization, and Preservation*, 5th edition, Chapter 8. Lippincott Williams & Wilkins, Philadelphia, pp. 159–183.
    (d) Compton W.D. 1989. A History of Apollo Lunar Exploration Missions, Diane Publishing Co, Darby, Pennsylvania, p146.

12. (a) Messier P.J. 2012. Facemask with filtering closure. US 8091551 B2.
    (b) Ohayon D., Bourget S., Gendron A.M., Tanelli J., Low K, Messier P.J., 2006. Triosyn technology on the breakdown of chemical warfare agents and industrial chemicals in the workplace. In: *The 9th Symposium on Iodine Utilization*, October 24, Chiba University, Chiba, Japan.
    (c) Bourget S., Ohayon D., Tanelli J., Gendron A.M., Messier P.J. 2004. Microbiocidal filtering media for individual protection: Facemasks and canisters. *8th International Symposium on Protection Against Chemical and Biological Warfare Agents*, Gothenburg, Sweden.
13. Inoue O., Yoshikawa M., Takaku M., Kaiho T., Taguchi M., Sambe H., Terada Y., Takaha T. 2015. Iodine and amylose containing fibers, method for production thereof, and use thereof. US 9206531 B2.
14. Hagiwara S. 2001. Cyclodextrin-iodine inclusion complex (CDI) as a new antibacterial agent. *The 4th Symposium on Iodine Utilization*, October 23, Chiba University, Chiba, Japan.
15. (a) https://pubchem.ncbi.nlm.nih.gov/compound/62097#section=Top (accessed March 23, 2017).
    (b) https://pubchem.ncbi.nlm.nih.gov/compound/1-__Diiodomethyl_sulfonyl_-4-methylbenzene#datasheet=lcss&section=Top (accessed March 23, 2017).
16. Kaiho T. 2014. Part V. Industrial applications of iodine: Other industrial applications: Colorant. In: Kaiho T. (ed.), *Iodine Chemistry and Applications*. John Wiley & Sons, Hoboken, NJ, pp. 552–553.
17. (a) IDD and their control, and global progress in their elimination. In: de Benoist B., Burrow G., Schultink W. 2007. *Assessment of Iodine Deficiency Disorders and Monitoring their Elimination: A Guide for Programme Managers*, 3rd ed. World Health Organization, Geneva, pp. 6–14.
    (b) Zimmermann M.B. 2009. Iodine deficiency. *Endocr. Rev.* 30:376.
    (c) Zimmermann M.B, Jooste P.L., Pandav C.S. 2008. Iodine-deficiency disorders. *Lancet* 372(9645):1251–1262.
18. Küpper F.C. et al. 2008. Iodide accumulation provides kelp with an inorganic antioxidant impacting atmospheric chemistry. *Proc. Natl. Acad. Sci. USA* 105(19):6954–6958.
19. Soda E., Kondo S., Ichihashi Y., Sato A., Ohtake H., Samukawa S., Saito S. 2007. Low-damage low-k etching by CF_3I plasma with low global warning potential. *American Vacuum Society 54th International Symposium & Exhibition*, October 16, PS1-TuA8, Seattle.
20. Kaiho T. 2006. Production and applications of iodine. *The 2nd International Conference on Hypervalent Iodine Proceedings*, June 1–2, Thessaloniki, Greece.
21. Kunimasa S., Tasaki M., Yamada H., Morimoto S., Togai M. *Sumitomo Kagaku*, Vol. 2012, Sumitomo R&D report, pp. 1–10. https://www.sumitomo-chem.co.jp/english/rd/report/theses/docs/2012E_3.pdf (accessed December 12, 2016).
22. (a) Chesterfield R. et al. Organic electronic devices. US 20080061685 A1.
    (b) Yamasaki Y. et al. Process for preparing triarylamine dimer. US 6242648 B1.
    (c) Qiu Y., Qiao J. 2000. Photostability and morphological stability of hole transporting materials used in organic electroluminescence. *Thin Solid Films* 372(1):265–270.
23. (a) http://www.fujifilm.com/innovation/technologies/control-of-light/ (accessed December 12, 2016).
    (b) Kawata K. 2002. Orientation control and fixation of discotic liquid crystal. *Chem. Rec.* 2:59–80.

(c) Negoro M., Yamaguchi J., Kawata K. Optical compensatory sheet and liquid crystal display. US6380996 B1.

24. Goh M., Matsushita S. Akagi K. 2010. From helical polyacetylene to helical graphite—Synthesis in chiral nematic liquid crystal field and morphology-retaining carbonisation. *Chem. Soc. Rev.* 39:2466–2476.

25. http://chemistry.elmhurst.edu/vchembook/548starchiodine.html (accessed December 12, 2016).

26. (a) Concise International Chemical Assessment Document 72 Iodine and Inorganic Iodides: Human Health Aspects, First draft prepared by John F. Risher and L. Samuel Keith, United States Agency for Toxic Substances and Disease Registry (ATSDR), Atlanta, Georgia, USA ©World Health Organization 2009.
(b) Shelor C.P., Dasgupta P.K. 2011. Review of analytical methods for the quantification of iodine in complex matrices. *Anal. Chim. Acta* 702:16–36.
(c) Hanna iodine checker. http://hannainst.com/hi718-iodine.html (accessed December 12, 2016).

27. (a) https://www.shodex.com/en/dc/07/01/02.html (accessed December 12, 2016).
(b) https://www.shodex.com/en/dc/07/05/47.html (accessed December 12, 2016).

28. (a) Firestone D. 1994. Determination of the iodine value of oils and fats: Summary of collaborative study. *J. AOAC Int.* 77(3):674–676.
(b) Porter N.A., Caldwell S.E., Mills K.A. 1995. Mechanisms of free radical oxidation of unsaturated lipids. *Lipids*, 30: 277–290.

29. (a) Karl F. 1935. Neues Verfahren zur maßanalytischen Bestimmung des Wassergehaltes von Flüssigkeiten und festen Körpern. *Angew. Chem.* 48(26): 394–396.
(b) http://www.mcckf.com/english/what.html (accessed December 12, 2016).

30. (a) https://www.jaea.go.jp/english/04/ntokai/houkan/houkan_01.html (accessed December 12, 2016).
(b) http://web.stanford.edu/group/scintillators/scintillators.html (accessed December 12, 2016).
(c) https://en.wikipedia.org/wiki/Scintillation_counter (accessed December 12, 2016).

31. (a) http://www.horiba.com/scientific/products/raman-spectroscopy/raman-academy/raman-faqs/what-is-raman-spectroscopy/ (accessed April 15, 2017).
(b) Svensson P, Kloo L. 2003. Synthesis, structure, and bonding in polyiodide and metal iodide–iodine systems. *Chem. Rev.* 103:1649.
(c) Andersson A., Sun T.S. 1970. Raman spectra of molecular crystals I. Chlorine, bromine, and iodine. *Chem. Phys. Lett.* 6:611.

32. (a) http://www.spring8.or.jp/en/about_us/whats_sp8/ (accessed December 12, 2016).
(b) http://www2.kek.jp/imss/pf/eng/about/sr/ (accessed December 12, 2016).
(c) Yamaguchi N., Nakano M., Takamatsu R., Tanida H. 2010. Inorganic iodine incorporation into soil organic matter: Evidence from iodine K-edge x-ray absorption near-edge structure. *J. Environ. Radioact.* 101(6):451–457.

33. (a) Jones J.H. 2000. The CativaTM process for the manufacture of acetic acid. *Platinum Met. Rev.* 44(3):94–105.
(b) Sunley G.J., Watson D.J. 2000. High productivity methanol carbonylation catalysis using iridium—the CativaTM process for the manufacture of acetic acid. *Catal. Today* 58(4):293–307.

34. Kissa E. 2001. *Fluorinated Surfactants and Repellents*, 2nd ed. Marcel Dekker Inc., New York.

35. Shirakawa H., Hiroki K. Fundamentals of conductive polymers. In: *Material Matters Basics*, Vol. 8. Sigma-Aldrich, Japan, pp. 1–5.

36. (a) MacDonald B. 2011. *World PA6 & PA66 Supply/Demand Report 2011 by PCI Nylon*. PCI Nylon GmbH, Bad Homburg.
    (b) Zweifel H., Ralph M.D., Schiller M. 2009. *Plastic Additive Handbook*, 6th ed. Hanser Fachbuchverlag, Germany, p. 80.
    (c) Scott G. (ed.). 1993. *Atomspheric Oxidation & Antioxidants*, Vol. 2. Elsevier Science Publishers, Amsterdam, pp. 141–218.
    (d) Janssen K., Gijsman P., Tummers D. 1995. Mechanistic aspects of the stabilization of polyamides by combinations of metal and halogen salts. *Polym. Degrad. Stab.* 1995, 49 (1), 127–133.

37. Crivello J.V. 1999. The discovery and development of onium salt cationic photoinitiators. *J. Polym. Sci. Part A Polym. Chem.* 37(23):4241–4254.

38. (a) Nagasaki N., Suzuki N., Nakano S., Kunihiro N. 1999. Process for producing iodotrifluoromethane. US Patent 5892136.
    (b) Nagasaki N., Morikuni Y., Kawada K., Arai S., 2004. Study on a novel catalytic reaction and its mechanism for CF_3I synthesis. *Catalysis Today*, 88: 121–126.
    (c) Meyer D. 1999. Commercialization of CFJ for fire-extinguishing systems in normally unwanted areas. *Halon Options Technical Working Conference*, April 27–29, New Mexico Engineering Research Institute, Albuquerque, NM, pp. 211–221.
    (d) http://www.nedo.go.jp/content/100799098.pdf (accessed December 12, 2016).

39. (a) Kodama H. 1981. Automatic method for fabricating a three dimensional plastic model with photo hardening polymer. *Rev. Sci. Instrum.* 52:1770.
    (b) http://www.wako-chem.co.jp/kaseihin_en/WPI/index.htm (accessed December 12, 2016).

40. (a) Girard P., Namy J.L., Kagan H.B. 1980. Divalent lanthanide derivatives in organic synthesis. 1. Mild preparation of samarium iodide and ytterbium iodide and their use as reducing or coupling agents. *J. Am. Chem. Soc.* 102:2693–2698.
    (b) Edmonds D.J., Johnston D., Procter D.J. 2004. Samarium(II)-iodide-mediated cyclizations in natural product synthesis. *Chem. Rev.* 104(7):3371–3404.

41. (a) Dohi T., Fukushima K., Kamitanaka T., Morimoto K., Takenaga N., Kita Y. 2012. An excellent dual recycling strategy for the hypervalent iodine/nitroxyl radical mediated selective oxidation of alcohols to aldehydes and ketones. *Green Chem.* 14:1493–1501.
    (b) Togo H., Sakuratani K. 2002. Polymer-supported hypervalent iodine reagents. *Synlett.* (12):1966–1975.

42. Ouchi M., Terashima T., Sawamoto M. 2009. Transition metal-catalyzed living radical polymerization: Toward perfection in catalysis and precision polymer synthesis. *Chem. Rev.* 109(11):4963–5050.

43. Kaiho T. 2014. Part III. Synthesis of iodine compounds: Iodinating reagents. In: Kaiho T. (ed.), *Iodine Chemistry and Applications*. John Wiley & Sons, Hoboken, NJ, pp. 251–276.

44. Kaiho T. 2014. Part IV. Biological applications of iodine: Pharmaceuticals: Therapeutic agents. In: Kaiho T. (ed.), *Iodine Chemistry and Applications*. John Wiley & Sons, Hoboken, NJ, pp. 433–437.

45. Gottardi W., 2001. Iodine and iodine compounds. In: Block S.S. (ed.), *Disinfection, Sterilization, and Preservation*, 5th edition. Lippincott Williams & Wilkins, Philadelphia, pp. 159–183. Chapter 8.

46. (a) Satoskar R.S., Bhandarkar S.D., Rege N.N. 1973. *Pharmacology and Pharmacotherapeutics*, Vol. 1 Popular Prakashan Private Limited, Mumbai, India, p. 665.
    (b) http://www.emedmd.com/content/amoebic-infections (accessed December 12, 2016).

47. Krause W. 2002. *Contrast Agents II: Optical, Ultrasound, X-Ray and Radiopharmaceutical Imaging.* Springer-Verlag, Berlin.
48. (a) http://www.who.int/ionizing_radiation/pub_meet/tech_briefings/potassium_iodide/en/ (accessed December 12, 2016).
    (b) Sternthal E., Lipworth L., Stanley B., Abreau C., Fang S.L., Braverman L.E. 1980. Suppression of thyroid radioiodine uptake by various doses of stable iodide. *N. Engl. J. Med.* 303:1083–1088.
    (c) Yoshida S., Ojino M., Ozaki T., Hatanaka T., Nomura K., Ishii M., Koriyama K., Akashi M. 2014. Guidelines for iodine prophylaxis as a protective measure: Information for physicians. *JMAJ* 57(3):113–123.
49. (a) http://www.nmp.co.jp/eng/products/index.html.
    (b) Adak S., Bhalla R., Vijaya Raj K.K., Mandal S., Pickett R., Luthra S.K. 2012. Radiotracers for SPECT imaging: Current scenario and future prospects. *Radiochim. Acta* 100:95–107.
50. (a) Mallela V.S., Ilankumaran V., Rao N.S. 2004. Trends in cardiac pacemaker batteries. *Indian Pacing Electrophysiol J.* 4(4):201–212.
    (b) Julien C., Mauger A., Vijh A., Zaghib K. 2015. *Lithium Batteries: Science and Technology.* Springer, New York.
51. (a) Sundberg J., Meller R. 1997. A retrospective review of the use of cadexomer iodine in the treatment of chronic wounds. *Wounds* 9(3):68–86.
    (b) Mertz P.M., Oliveira-Gandia M.F., Davis S.C. 1999. The evaluation of a cadexomer iodine wound dressing on methicillin resistant *Staphylococcus aureus* (MRSA) in acute wounds. *Dermatol Surg.* 25(2):89–93.
52. Berg J.N., Maas J.P., Paterson J.A., Krause G.F., Davis L.E. 1984. Efficacy of Ethylenediamine Dihydriodide as an Agent to Prevent Experimentally Induced Bovine Foot Rot, *Am. J. Vet. Res.* 45(6):1073–1078.
53. (a) Borucki S., Berthiaume R., Robichaud A., Lacasse P. 2012. Effects of iodine intake and teat-dipping practices on milk iodine concentrations in dairy cows. *J. Dairy Sci.* 95(1):213–220.
    (b) Gleeson D., O'Brien B., Flynn J., O'Callaghan E., Galli F. 2009. Effect of pre-milking teat preparation procedures on the microbial count on teats prior to cluster application. *Ir Vet J.* 62(7):461–467.
54. (a) Fletcher W.W., Smith J.E. 1964. The growth of bacteria, fungi, and algae in the presence of 3,5-dihalogeno-4-hydroxybenzonitriles with comparative data for substituted aryloxyalkanecarboxylic acids. *Proceedings of the British Weed Control Conference 7th*, August 2–4, Glasgow, Vol. 1. British Crop Protection Council, England, p. 20.
    (b) Simmonds M.J. 1968. Experiments on weed control with hydroxybenzonitrile formulations in salad and bulb onions. *Proceedings of the British Weed Control Conference 9th*, November 18–21, Vol. 1. British Crop Protection Council, England, p. 344.
    (c) http://www.fao.org/fileadmin/templates/agphome/documents/Pests_Pesticides/Specs/Old_specs/IOOC.pdf.
    (d) Gutteridge S, Thompson M.E. 2012. Acetohydroxyacid synthase inhibitors (AHAS/ALS): Biochemistry of the target and resistance. In: Jeschke P., Krämer W., Schirmer U., Witschel M. (eds.), *Modern Methods in Crop Protection Research*, Wiley-VCH Verlag GmbH, Weinheim, p. 29.
    (e) Hacker E., Bieringer H., Willms L., Ort O., Koecher H., Kehne H. 1999. Iodosulfuron plus mefenpyr-diethyl—A new foliar herbicide for weed control in cereals. *Proceedings of the Brighton Crop Protection Conference, Weeds*, November 15–18, Vol. 1, Brighton, UK. p. 15.

151

55. (a) Selby T.P., Sternberg C.G., Bereznak J.F., Coats R.A., Marshall E.A. 2007. The discovery of proquinazid: A new and potent powdery mildew control agent. In: Lyga J.W., Theodoridis G. (eds.), *Synthesis and Chemistry of Agrochemicals VII.* Oxford University Press, Washington, DC, p. 209.
    (b) Hansen J. 1984. IPBC—A new fungicide for wood protection. *Mod. Paint Coat.* 74:50, 52, 55, 90.
    (c) Frost A.J.P., Hampel M. 1976. The development of benodanil for the control of cereal rust diseases. *Proceedings of the European and Mediterranean Cereal Rusts Conference 4th, Interlaken.* Swiss Fed. Res. Stn. Agron. Zurich, Switzerland, p. 99.
56. (a) Tohnishi M., Nakao H., Furuya T., Seo A., Kodama H., Tsubata K., Fujioka S., Kodama H., Hirooka T., Nishimatsu T. 2005. Flubendiamide, a novel insecticide highly active against lepidopterous insect pests. *J. Pestic. Sci.* 30(4):354–360.
    (b) Ebbinghaus-Kintscher U., Luemmen P., Lobitz N., Schulte T., Funke C., Fischer R., Masaki T., Yasokawa N., Tohnishi M. 2006. Phthalic acid diamides activate ryanodine-sensitive Ca^{2+} release channels in insects. *Cell Calcium* 39:21–33.
    (c) Masaki T., Yasokawa N., Tohnishi M., Nishimatsu T., Tsubata K., Inoue K., Motoba K., Hirooka T. 2006. Flubendiamide, a novel Ca^{2+} channel modulator, reveals evidence for functional cooperation between Ca^{2+} pumps and Ca^{2+} release. *Mol. Pharmacol.* 65:1733–1739.
57. http://www.ioteq.com/ (accessed December 12, 2016).
58. (a) Vonnegut B., Chessin H. 1971. Ice Nucleation by Coprecipitated Silver Iodide and Silver Bromide. *Science.* 174 (4012): 945–946.
    (b) Cloud-seeding brings rain to Tama. *The Japan Times News*, August 22, 2013.
    (c) Phyllis A. Lyday 2005. *Iodine and Iodine Compounds, Ullmann's Encyclopedia of Industrial Chemistry*, Wiley-VCH, Weinheim, Section 7 Uses.
59. (a) O'Regan B., Grätzel M. 1991. A low-cost, high-efficiency solar cell based on dye-sensitized colloidal TiO2 films. *Nature* 353(6346):737–740.
    (b) Hara K., Arakawa H. 2005. Dye-sensitized solar cells, Chapter 15. In: Luque A., Hegedus S. (eds.), *Handbook of Photovoltaic Science and Engineering.* John Wiley & Sons, Hoboken, NJ, pp. 663–700.
    (c) Press Release of EPFL http://actu.epfl.ch/news/dye-sensitized-solar-cells-rival-conventional-ce-2/ (accessed December 12, 2016).
60. (a) IS Process Research & Development Group, JAEA https://www.jaea.go.jp/04/o-arai/nhc/en/intro/is/is_top.htm (accessed December 12, 2016).
    (b) Noguchi H. et al. 2014. Components development for sulfuric acid processing in the IS process. *Nucl. Eng. Des.* 271:201–205.
    (c) Tanaka N., Yamaki T., Asano M., Maekawa Y., Terai T., Onuki K. 2012. Effect of temperature on electro-electrodialysis of $HI-I_2-H_2O$ mixture using ion exchange membranes. *J. Membr. Sci.* 411–412:99–108.
    (d) Kubo S., Tanaka N., Iwatsuki J., Kasahara S., Imai Y., Noguchi H., Onuki K. 2012. R&D status on thermochemical IS process for hydrogen production at JAEA. *Energy Procedia* 29:308–317.
61. (a) Advanced NASA Technology Supports Water Purification Efforts Worldwide. February 29, 2012.
    (b) https://www.nasa.gov/mission_pages/station/research/benefits/water_purification.html (accessed December 12, 2016).
62. (a) Johansson Seechurn C.C.C., DeAngelis A., Colacot T.J. 2014. *New Trends in Cross-Coupling: Theory and Applications*, RSC Catalysis Series, Royal Society of Chemistry, Cambridge, UK, pp. 1–19.

(b) Yamaoka N., Sumida K., Itani I., Kubo H., Ohnishi Y., Sekiguchi S., Dohi T., Kita Y. 2013. Single-electron-transfer (SET)-induced oxidative biaryl coupling by polyalkoxybenzene-derived diaryliodonium(III) salts. *Chem.—Eur. J.* 19(44):15004–15011.

(c) Zhdankin V.V. 2013. *Hypervalent Iodine Chemistry: Preparation, Structure, and Synthetic Applications of Polyvalent Iodine Compounds.* John Wiley & Sons, Inc., Hoboken, NJ.

(d) Hypervalent Iodine Compounds http://www.tcichemicals.com/eshop/en/jp/category_index/03037/ (accessed December 12, 2016).

63. (a) Endo M., Osaka T., Takeda S. 2004. High-efficiency chemical oxygen–iodine laser using a streamwise vortex generator. *Appl. Phys. Lett.* 84:2983.

(b) Endo M. 2008. New iodine laser achieves positive gain. *SPIE Newsroom* Article 26878 (August 11).

(c) Endo M., Walter R.F. (eds.). 2006. *Gas Lasers.* CRC Press, Boca Raton, FL, December 26.

64. (a) *1st International Symposium on Halogen Bonding (ISXB-1)*, June 18, 2014. http://www.isxb-1.eu/ (accessed December 12, 2016).

(b) Desiraju G.R., Ho P.S., Kloo L., Legon A.C., Marquardt R., Metrangolo P., Politzer P., Resnati G., Rissanen K. 2013. Definition of the halogen bond (IUPAC Recommendations 2013). *Pure Appl. Chem.* 85(8):1711–1713.

(c) Imakubo T., Murayama R. 2013. Solvent dependence of crystal morphology, donor/anion ratio and electrical conductivity in a series of hexagonal supramolecular organic conductors based on diiodo(pyrazino)tetrathiafulvalene (DIP). *Cryst. Eng. Comm.* 15:3072–3075.

(d) Scholfield M.R., Vander Zanden C.M., Carter M., Ho P.S. 2013. Halogen bonding (X-bonding): A biological perspective. *Protein Sci.* 22:139–152.

(e) Roughley S.D., Jordan A.M. 2011. The medicinal chemist's toolbox: An analysis of reactions used in the pursuit of drug candidates. *J. Med. Chem.* 54:3451.

65. Wang L., Zhou X., Fredimoses M., Liao S., Liu Y. 2014. Naturally occurring organoiodines. *RSC Adv.* 4:57350–57376.

66. Column 1. http://seihou8.sakura.ne.jp/shop/008-jinnou/index-l.htmls

67. Column 2. Kahr B., Freudenthal J., Phillips S., Kaminsky W. 2009. Herapathite. *Science* 324:5933, 1407.

68. Column 3. (a) Tonacchera M. et al. 2013. Iodine Fortification of Vegetables Improves Human Iodine Nutrition: In Vivo Evidence for a New Model of Iodine Prophylaxis. *J. Clin. Endocrinol. Metab.* 98(4):E694–E697.

(b) Kiferle C, Gonzali S., Holwerda H.T., Ibaceta R.R., Perata P. 2013. Tomato fruits: a good target for iodine biofortification. *Front Plant Sci.* 4:205.

69. Column 4. http://medical.nikkeibp.co.jp/inc/all/hotnews/archives/282267.html

70. Column 5. Carmeli D. 1991. Method of detecting counterfeit paper currency. US 5063163 A.

71. Column 6. Lee M. D., Manning J. K., Williams D. R., Kuck N. A., Testa R. T., Borders D. B. 1989. Calichemicins, a novel family of antitumor antibiotics. *J. Antibiotics* 42 (7): 1070–87.

72. Column 7. Nanjyo M. 1998. Urban Mine, New Resource for the Year 2000 and Beyond. Bulletin of the Research Institute of Mineral Dressing and Metallurgy, Tohoku University, 1998, 43: 239–251.

73. Column 8. Kojima A., Teshima K., Shirai Y., Miyasaka T. 2009. Organometal Halide Perovskites as Visible-Light Sensitizers for Photovoltaic Cells. *J. Am. Chem. Soc.* 131(17):6050–6051.

153

Index

159